Measurement
Analysis

Measurement Analysis

An Introduction to the Statistical Analysis of Laboratory Data in Physics, Chemistry and the Life Sciences

Mario Caria

University of Cagliari, Italy

Imperial College Press

Published by

Imperial College Press
57 Shelton Street
Covent Garden
London WC2H 9HE

Distributed by

World Scientific Publishing Co. Pte. Ltd.
P O Box 128, Farrer Road, Singapore 912805
USA office: Suite 1B, 1060 Main Street, River Edge, NJ 07661
UK office: 57 Shelton Street, Covent Garden, London WC2H 9HE

British Library Cataloguing-in-Publication Data
A catalogue record for this book is available from the British Library.

ISBN 1-86094-231-8

Printed in Singapore by World Scientific Printers

Contents

to Clemente, who inspired the desire to sit down and write.
to Teodolinda, who gave the strength to write it all.
to Ulderico, who brought good fortune of having it published.

1 Measurements, errors and estimates

1.1 Error and Uncertainty

In general we can say that we all agree more or less on the meaning of the word *error*. Error gives indications on experimental results and the quality of measurements. It thus helps to arrive at conclusions on experimental data and discern their reliability.

The experimentalist knows, starting from design phase, that his experiment will not be perfect despite all his efforts. He/she knows that the ideal *perfect* experiment will not be matched by the *real* one. Everyone knows that measurement is no more than an approximation of the ideal. To quantify this approximation one must employ common tools in describing experimental results, their quality and considerations on them as unambiguously and accessibly as possible. These tools, which are introduced here, are an essential part of the experimentalist's education.

Let us introduce a definition of error that best follows our *common sense* as 'doubt about a measurement' and let us make sure that others share this definition and, on this basis, draw the same conclusions.

The concept of error is associated, in the definition given in recent *I.S.O.* [1] rules, with the more general term of *uncertainty*, with which the experimental result is evaluated and with which it is associated.

As concerns the scientific disciplines, we take for granted that the best evaluation of an experimental result is the numerical one [2], which simplifies synthesis and facilitates the communication of results. We therefore consider this also as a criterion in evaluating the *uncertainty* we wish to communicate. In what follows we will try to give a numerical evaluation of *uncertainty* associated with experimental results. However, it is often stated that one wishes to determine the *accuracy* of

---

[2]This is not necessarily true but, in a modern teaching approach common to the great majority of teachers and researchers, is taken for granted.

the experimental result [3]. Contrary to what one may believe, it is not true that the only information one has is the result itself. The experimentalist also applies his/her own judgement in evaluating the validity of a measurement. This is the reason why the concept of uncertainty, which is a general one, must be kept separate, including in the numerical expression the different origins that lead to a certain error.

In what follows we will schematically divide the possible causes of errors and present an evaluation of the corresponding uncertainties which can be encountered, at least in most experimental situations. We will do this more for the sake of a common terminology than to limit the experimentalist's judgement.

1.1.1   Resolution and reading error

We can distinguish a type of error which we make, for example, <u>once</u> on measuring a single time interval with a single measurement. We suppose that this measurement has been taken using one of our sensorial organs: the eye. As an example, we can use the reading of an instrument in a certain interval of time.

This definition of measurement may seem arbitrary, but actually it is quite rigorous and points up arbitrariness in the measurement, both in the choice of the sensorial organs (the kind of experimental detection) and of the interval of time.

Let us try to understand through an example. Let us suppose we want to read temperature by means of a thermometer. We do this by observing marks associated with numerical values corresponding to the height of the mercury column. In reality, we are making a measurement

---

[3] *"Accuracy"* and *"precision"* are often confused, but the concepts are quite different depending upon the contexts in which they are used. Several definitions of *precision* can be found in the references. This term will not be used here. By *accuracy* we mean the qualitative evaluation, as presented here, of the validity of the measurement. Therefore, great accuracy signifies small uncertainty and vice versa. It is not, however, its opposite, since the latter will be given a precise numerical value.

with the eyes and we are choosing a reasonable interval of time. We commonly use the blink of the eye, which is more or less one second. If we chose a longer interval, for example 1000 seconds, we know very well that the measurement would be different. On the other hand, if we choose to determine the height of a person with a yardstick, we know very well that the measurement will give us the same result whether it is made during an interval of one second or during an interval of 1000 seconds. This example shows that we cannot define experimental measurement in a general way, but must relate the definition to the single experimental situation. To arrive at a general criterion we can base our reading error on the specifications of the instrument. The manufacturer provides the definition of the instrument's characteristics by giving the measurement resolution [4].

If the measurements were just as *accurate*, the resolution or uncertainty would correspond exactly with the *reading error*.

Let us go back to the example. The height of the mercury column has to correspond to a certain mark on the ruler. The manufacturer's mark corresponds, for example, to one degree. From that we cannot infer that the resolution of the instrument, namely the reading error, is 1 °C. This would be wrong, or in any case arbitrary. Very often, in fact, the resolution is not indicated in the technical specifications. Even when this is indicated, what we are about to say is still valid.

Let us consider two different thermometers, one small enough to hold in our hand, the other two meters high. Let us suppose they have two identical scales from one end to the other. By identical scales we really mean that they have the same value in degrees°C from the beginning to the end of the scale (for example from 30°C to 40°C) and that they have the same number of marks (eleven). They will have the same mercury bulb at the base, of course of different size. If we follow the same line of reasoning as before, they should give measurements with

---

[4]In reality, what is given is an *intrinsic error* or a *fiducial error*, determined with respect to standard reference values.

the same accuracy. However, on the larger thermometer two consecutive marks are much further apart (without even quantifying, it is enough to consider the fact that there are several centimeters between them) than the marks on the smaller one. The eye can distinguish positions of the mercury column between two marks on the larger thermometer that it cannot evaluate on the smaller one. Experimentalists may thus be induced to determine the resolution of the instrument as being greater than the distance between two marks. The reading error, as well as the measurement, would not have a general definition. On the contrary, it would be arbitrarily determined by each experimentalist in a different manner. In fact, this is the case. Here we can only give commonsense indications, which are discussed at length in chapter **??**.

When reading instruments, the marker (for example the arrow of a scale or, as in this case, the top of the mercury column) should be much smaller than the interval between two consecutive marks. If this is not the case, evaluation of uncertainty, which is to say estimating the reading error, must be performed by taking the inter-distance between two consecutive marks. Depending on to what extent this is verified, we can estimate the resolution as equal to the interval between two marks divided by two, three, four, and so on. It is clear, however, that the manufacturer will put appropriate, and not arbitrary, marks on the instrument. We do not consider reasonable, and experience supports us in this, a presumed resolution larger than the inter-distance between the marks divided by four.

These are indications of a wide consensus in the evaluation of the reading error of an instrument in the case of experimental measurements made by the sensorial organ of the eyes [5].

In what follows we will find also a more rigorous justification of the estimation of one half or of one quarter of the interval between two

---

[5]In the case of measurements taken by other sensorial organs, like the ear, it is more complicated. Suffice it here to indicate that measurements taken by the sensorial organ of the ear are far more arbitrary [2].

consecutive marks.

For the moment let us go on with the discussion of error now that we have a definition of the estimate of uncertainty due to the reading error, which will be used for the definition of casual error.

1.1.2   The illegitimate error

While measuring, we may make wrong assumptions, mistakes in calculating or use the wrong units. We can call this kind of error *illegitimate*. Illegitimate errors can later be corrected by more careful analysis. Here we can also include all possible operations that allow us to make better measurements by improving environmental conditions or using better instruments. In the following we shall give an example of the difficulty in defining illegitimate error.

Example: the problem of meter positioning in the experiment with coaxial cables

In the case of the experiment for the determination of the speed of light in coaxial cables (described in the appendix), we need to measure an electrical cable with a meter bar that is slightly shorter than the cable. One tries to place one end of the cable exactly at the end of the metre bar. But then, since the cable is longer, one has to move the meter bar. During this operation the cable often bends owing to its normal elastic properties. One therefore has to put one finger on the cable at the one-metre mark and at the same time keep the two parallel - a difficult operation, especially if one is using a carpenter's metre. To complete the measurement one then has to position the end of the metre once again and measure the remaining part (uncertainties must be added up, as discussed in the following chapters).

In the first reading it is legitimate to consider just the reading error of the meter, which corresponds to the smallest mark on the scale, let us say, 0.1 centimetres. But in the second measurement, if we take the same reading error, which is later to be added to the first to have the full uncertainty of the measurement over the entire length of the cable, we could argue that the smallest mark, or better the smallest interval between two marks, underestimates the error. This is due to the fact that positioning for the second measurement is not as accurate as it was for the first. One student pointed out that the error could be estimated as equal to the width of the thumb with which he marked the end of the first measurement of the cable and positioned it for the second.

This case is typical of an illegitimate error which experimentalists have to keep to a minimum. It is a question of the experimentalist's properly positioning the cable and not allowing it to bend, and the meter. One could also mark the end of the first measurement on the cable with a pen, for instance, which would certainly be more accurate than the thumb.

Similar to this are reading errors on a scale due merely to incorrect positioning of the reading plane with respect to the scale plane [6]

1.1.3   Uncertainty and systematic error

By this we mean something with less defined contours, which are common to many different experimental situations. This is due to the very nature of this kind of error. The systematic error must be distinguished from the illegitimate error because it is quite difficult to know in advance if we can avoid it. For example, one can call systematic error the one due to the uncalibrated scale of a thermometer or meter bar. If the manufacturer of a meter makes a mistake in numbering a scale, e.g. going directly from 10 to 30 omitting 20, one then all measurements

---

[6]This is very often called *parallax error*, because the axis orthogonal to the scale plane is not orthogonal to the plane of our eyes. But in this case the definition refers to the orthogonal axes and not to the planes themselves. This is very often classified among the systematic errors, which have a completely different meaning in this book.

above 10 will contain a systematic error of 10 cm. *Systematically* or *repeatedly* each reading will contain an error that can be evaluated as a *systematic uncertainty* equal to 10 cm. If we notice the defect after taking all the measurements, we can evaluate this defect numerically and make allowances for it in the experimental results. In this case we call it *illegitimate error*.

Experimentalists may also fail to realise that there are illegitimate errors or may have introduced more subtle measurement defects or ones generated by themselves. In this case they can only compare their measurements with those made by others *with different instruments*. In this way, on noting a discrepancy they can attribute to the measurement an evaluation of the systematic error, *in all compared measurements and for all measurements of all other experimentalists*.

1.2   Uncertainty and casual error

The conceptual distinction between the errors presented above is that *the casual error* [7] cannot be cancelled, neglected or evaluated a posteriori. This is true even if we reduce it so much as to be far smaller than the experimental result, which may occur in one of the two previous cases of errors. This is precisely the error that we have to evaluate and associate with the measurement since we are assuming that we can-

------

[7] As defined in most textbooks, it would be more appropriate to term the casual error the *probable*. This is also the name with an analogous definition given in reference [1]. Here it would be more suitable not to fault the definition but the name to ascribe to it for teaching purposes. Students are often confused by the word *casual*; they ascribe to it a meaning that is not synonymous with *accidental* and include in it the concept of systematic error and illegitimate error. By *probable* error one often means a quantity corresponding to an interval of values. In this textbook we discuss this quantity again in the last chapter. But we do not adopt its definition in common use (also not adopted in *I.S.O.*) regulation) nor its use.

l_1	8.0cm
l_2	8.1cm
l_3	7.9cm
l_4	8.1cm
l_5	8.2cm

Table 1.1: Values of five measurements of length of a pencil.

not have *perfect* measurements in any case [8]. By definition we can have only an estimate of it. If we knew the exact error, we would know the result itself with absolute accuracy and therefore it would not be necessary to evaluate the error. This would contradict what we said earlier on. It is the estimate which can be quantified in a rigorously mathematical way. Here we call the estimate of this error *casual uncertainty*. The methods used to obtain this estimate are the basis of the statistical analysis of experimental data. To obtain it, in fact, concepts and methods of mathematical analysis and probability calculations are required.

A first, easily intuitive and commonly used method can be deduced by a simple estimate of the numerical result itself by determining the arithmetical mean.

Let us consider the measurements in table 1.1, of lengths l_i of a common object such as a partly used pencil.

With $n = 5$ measurements, one can obtain the arithmetical mean of the measurements of the length, which is indicated as $\bar{l}$:

$$\bar{l} = \frac{\sum_{i=1}^{n} li}{n} = \frac{(8.0 + 8.1 + 7.9 + 8.1 + 8.2)cm}{5} = \tag{1.1}$$

$$= \frac{40.3cm}{5} \cong 8.06cm \cong 8.1$$

---

[8]This is taken for granted here, but in reality there is experimental evidence that this is really the case. It will be verified later on.

l_1	8.0cm	l_{11}	8.0cm	l_{21}	8.0cm
l_2	8.1cm	l_{12}	8.0cm	l_{22}	8.0cm
l_3	7.9cm	l_{13}	8.1cm	l_{23}	7.8cm
l_4	8.1cm	l_{14}	8.0cm	l_{24}	8.0cm
l_5	8.2cm	l_{15}	7.8cm	l_{25}	8.0cm
l_6	8.2cm	l_{16}	7.7cm		
l_7	8.0cm	l_{17}	8.0cm		
l_8	8.2cm	l_{18}	8.0cm		
l_9	8.1cm	l_{19}	8.1cm		
l_{10}	8.0cm	l_{20}	8.0cm		

Table 1.2: Values of 25 measurements of length of a pencil.

By making a much larger number of measurements, for example by having 24 other colleagues measure it, we can expect to obtain data of the kind in table 1.2.

We therefore have

$$\bar{l} = \frac{\sum_{i=1}^{n} li}{n} = \frac{\sum_{i=1}^{25} li}{25} = \frac{200.4cm}{25} \cong 8.0cm \tag{1.2}$$

This is an *estimate* of the value of length of the pencil.

We will come back(chapter 3.2) to this problem of significant digits. The digits of equations 1.1 and 1.2 are correctly evaluated, but they are not justified.

It is clear that the experimentalist knows that the pencil has only one length [9]. This is what is called *true value* or *exact value* obtained by a conceptual extrapolation and mathematically defined. We can then

---

[9]It is incorrect to write in this way, but to give full explanation it is necessary to introduce concepts of quantum mechanics, which the students may not be prepared for.

state that the arithmetical mean of equation 1.2 is the estimate of the true value. Actually, this is not exactly the case, and we will see later on how we can improve on this definition.

We started our discussion on the determination of uncertainty by assuming that we can make an estimate of it, as well as of the real value. We will therefore have to make an idealisation both of the measurement and of the error.

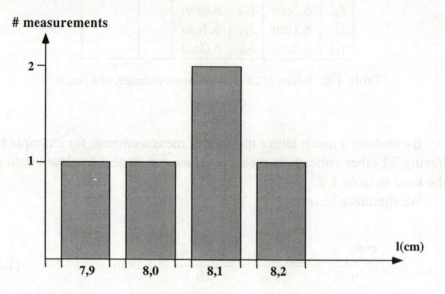

Figure 1.1: Histogram for five measurements of length of a pencil.

Let us suppose we have made an even larger number of measurements. For example, by having 24 other colleagues measure the pencil, just *"to be more certain"* of the result. In this case we are assuming that by increasing the number of measurements used in the determination of the arithmetic mean, the result should be more accurate. To quantify this statement, we are therefore stating that the error we are making should be smaller and therefore the uncertainty associated with the measurement should also be smaller.

Before we continue improving the definition, we see that by increasing the number of measurements, the arithmetical mean becomes a progressively better estimate of the true value of a quantity [10].

To show that, it is useful to write the values in a simple plot. In the abscissa we will put the values of the measurements and in the ordinates the values of the number of times this value has been obtained by measurement [11]. This plot is called a *histogram*. Figure 1.1 illustrates the distribution of data in table 1.1.

By comparing this figure with figure 1.2, for table 1.2 it is clear that measurements tend to assume a value that approaches the value obtained with the arithmetic mean.

We can now implement an idealised reasoning process. By making an ever-increasing number of measurements, we will obtain a progressively more accurate measurement, which is to say a better estimate of the true value, from the arithmetic mean. For an infinite number of measurements, we obtain a true value, which is the best estimate of itself, and which is the true value itself. By translating this reasoning into mathematical terms, we can write that the true value is :

$$\mu = \lim_{n \to \infty} \bar{l} = \lim_{n \to \infty} \left(\sum_{i=1}^{n} \frac{l_i}{n} \right) \tag{1.3}$$

In figure 1.3 the curve we see the curve of the *parent distribution* of the *experimental distribution* of figure 1.2.

The curve will not necessarily have the same shape as figure 1.3, it may also be asymmetric. A parent distribution can be associated with

---

[10] In reality we will see some cases in which this is not true. This is due to the fact that the whole reasoning has been built on the assumption that we now have to demonstrate. Thus, for the moment we do not need further indications but we are simply introducing intuitive concepts and defining them in a *more rigorous* manner.

[11] This is very often called *frequency*, but in statistical analysis this name usually indicates a different quantity. Here we therefore avoid this use so as to avoid ambiguities.

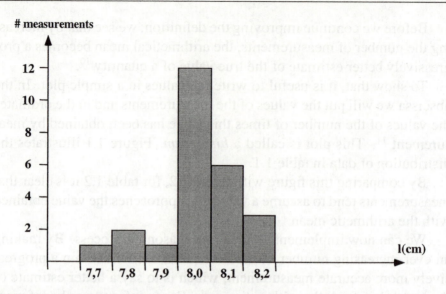

Figure 1.2: Histogram for 25 measurements of length of the same pencil used in the previous figure. This is an *experimental distribution*.

any experimental distribution. *Mean value* is the value of the abscissa obtained from equation 1.3, which ideally corresponds to the arithmetic mean. A warning note is that not only will the curve not always have a single and symmetric bell shape, but in general, the parent distribution will be difficult to identify from the experimental distribution. The limit will not have a possible numerical equivalence in practical terms. Therefore it is difficult to obtain the true value. For this reason we can use methods based on the concept of *maximum likelihood*. This will be the subject of chapter 4. Here we wish only to limit the discussion to arrive at a better understanding of the definition.

For example, in the curve in figure 1.4, the mean value is neither the one to which the maximum value of the ordinates corresponds, nor the one around which (to the right and to the left) the curve is symmetric in the values of the integral. The latter points are called *most probable*

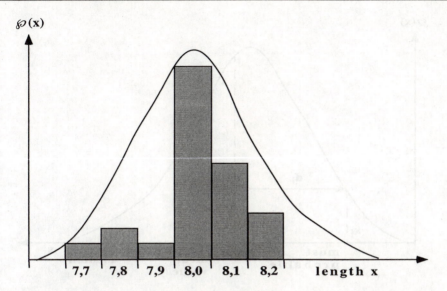

Figure 1.3: Parent disribution curve from the limit procedure, corresponding to the experimental distribution of the previous figure.

value or *mode* and *median* respectively. In the following we will justify this terminology. Let us go back to uncertainty.

We are following a line of reasoning to obtain a quantitative evaluation of the casual uncertainty.

In the curve in figure 1.4 we indicate the *deviations* of each value in abscissa, which means the differences between the value of the abscissa and the mean value:

$$d_i = x_i - \mu \tag{1.4}$$

It is clear that the only deviation equal to zero is that of the mean value. This statement may be obvious but it is best to keep in mind in the following discussion.

For the deviations, we can implement a limit procedure as well. Even though the d_i are as many as the l_i with $i = 1, \ldots, \infty$ we can

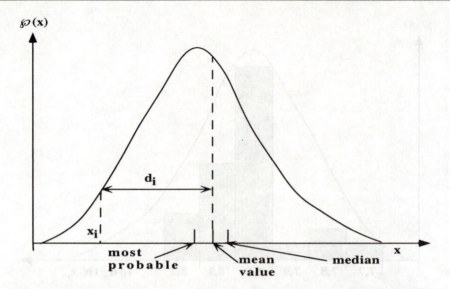

Figure 1.4: Curve of the parent distribution corresponding to the previous experimental distribution. The shape is asymmetric; it is possible to define the characteristic parameters, *mean value*, *median*, *most probable value* , and the *deviations*.

calculate the arithmetical mean of a subsample of n measurements and calculate the deviations d_i with $i = 1, \ldots , n$, therefore the arithmetical mean of the deviations themselves, obtaining the mean deviation $\overline{d}$:

$$\overline{d} = \frac{\sum_{i=1}^{n} d_i}{n} \tag{1.5}$$

Going to the limit for the deviations similar to what we have done for the l_i:

$$\lim_{n\to\infty} \overline{d} = \lim_{n\to\infty} \left(\sum_{i=1}^{n} \frac{d_i}{n}\right) = \lim_{n\to\infty} \left(\sum_{i=1}^{n} \left(\frac{l_i - \mu}{n}\right)\right) = \tag{1.6}$$

$$= \lim_{n\to\infty} \left(\frac{1}{n}\left(\sum_{i=1}^{n}(l_i - \mu)\right)\right) = \lim_{n\to\infty} \left(\frac{1}{n}\left(\sum_{i=1}^{n}(l_i) - \frac{1}{n}\left(\sum_{i=1}^{n}(\mu)\right)\right)\right) =$$

$$= \lim_{n\to\infty} \left(\frac{1}{n}\right)\sum_{i=1}^{n}(l_i) - \lim_{n\to\infty} \left(\frac{1}{n}\right)\left(\sum_{i=1}^{n}\mu\right) =$$

$$= \lim_{n\to\infty} \left(\frac{1}{n}\right)\sum_{i=1}^{n}(l_i) - (\mu) = \mu - \mu = 0$$

With what we said before the deviations defined in this way are equal to zero for the mean value. The next steps in arriving at equation 1.2 are needed to state that the d_i are *the best estimate of the error*. In fact, their mean value is equal to zero at ∞, namely when the arithmetical mean of the measurements reaches the true value. This is exactly what we wished to state. We have, in fact, found a correspondence between the idealisation of being able to find the true value and the fact that in this case we do not make *any error*.

Thus the uncertainty for each measurement is obtained with a limit procedure on the deviations.

To maintain a definition of an always positive quantity, the common practice is to consider the modulus of the deviations, in the following way [11]:

---

[11]To continue the discussion we introduce here the definitions of modulus and squares of the deviations, which may seem arbitrary. In reality, they are introduced simply because they keep the meaning of deviation from the true value and they do not equal zero at ∞.

$$\eta = \lim_{n \to \infty} \left(\sum_{i=1}^{n} \left(\frac{|l_i - \mu|}{n} \right) \right) > 0 \tag{1.7}$$

It is common use to define the square of same quantity in the summation.

We can now reconsider the limit of equation 1.5 which is called *variance* [12]:

$$\xi = \lim_{n \to \infty} \left(\sum_{i=1}^{n} \left(\frac{|l_i - \mu|^2}{n} \right) \right) \tag{1.8}$$

To have dimensions comparable with the measured quantity, it is advisable to consider the square root of this quantity:

$$\sigma = \sqrt{\xi} = \sqrt{\lim_{n \to \infty} \left(\sum_{i=1}^{n} \left(\frac{|l_i - \mu|^2}{n} \right) \right)} \tag{1.9}$$

which is called *standard deviation of the parent distribution*.

We judge the standard deviation as the best estimate to associate with uncertainty due to the casual error. We usually tend to *identify standard deviation with casual uncertainty*.

It is still necessary to find a quantity that we can use in a correct quantification of the uncertainty to associate with the measurement.

In practice, it is not easy to calculate the true value. Therefore in the definition of equation 1.9 we replace the mean value with the arithmetical mean with approximation:

$$\xi \cong \omega = \sqrt{\sum_{i=1}^{n} \left(\frac{|l_i - \bar{l}|^2}{n} \right)} \tag{1.10}$$

---

[12]This terminology is of more general use in statistics. Variance must be specified with respect to the intended quantity. In our case it will always mean variance with respect to the mean value

Going back to the previous observation, that the only deviation equal to zero is that of the mean value, it is obvious that we cannot calculate the deviation of the mean if we include the mean value itself in the division. In fact this contributes to the deviation as zero and would therefore lead to underestimation of the result of the mean deviation. To avoid this contradiction [13] in the redefinition of the standard deviation, we will introduce the $n - 1$ factor in the calculation instead of n. It is to be noted that, in any case, in the approximation for $n \to \infty$, the difference is negligible. Therefore the equation 1.10 becomes:

$$\xi \cong s = \sqrt{\sum_{i=1}^{n} (\frac{|l_i - \bar{l}|^2}{n - 1})} \tag{1.11}$$

Thus this is the quantity considered as being the *standard deviation of the experimental distribution* or simply *standard deviation*. We identify the estimate of the casual uncertainty with the experimental standard deviation which will be used from now on.

Going back to the example of table 1.2, and considering that all measurements were performed with the same instrument having resolution 0.1cm.

In the case of the table the standard deviation, is :

$$s = \sqrt{(\sum_{i=1}^{n} (\frac{|d_i|^2}{n - 1})} = \sqrt{\frac{0.36}{24}} \cong \sqrt{0.02} \cong 0.12 \cong 0.1cm \tag{1.12}$$

By repeating the same exercise for table 1.1 we obtain $s \cong 0.2cm$. This result justifies what was said above.

---

[13] Although this reasoning could have been introduced in the definition of the parent standard deviation, the author feels that it is easier to understand at this point

d_1	$0.0cm$	d_{11}	$0.0cm$	d_{21}	$0.0cm$
d_2	$0.1cm$	d_{12}	$0.0cm$	d_{22}	$0.0cm$
d_3	$0.1cm$	d_{13}	$0.1cm$	d_{23}	$0.2cm$
d_4	$0.2cm$	d_{14}	$0.0cm$	d_{24}	$0.0cm$
d_5	$0.2cm$	d_{15}	$0.2cm$	d_{25}	$0.0cm$
d_6	$0.2cm$	d_{16}	$0.3cm$		
d_7	$0.0cm$	d_{17}	$0.0cm$		
d_8	$0.2cm$	d_{18}	$0.0cm$		
d_9	$0.1cm$	d_{19}	$0.1cm$		
d_{10}	$0.1cm$	d_{20}	$0.0cm$		

Table 1.3: Values of the deviations for the 25 measurements of length of a pencil.

1.2.1   The expression of expanded uncertainty

Here we wish to give a *global* or *extended*, expression of uncertainty that includes the various types of error introduced and that allows their numerical evaluation in a brief and common notation, with which we can accompany the experimental result.

To obtain this we must discuss the combination of uncertainties and thus determination of combined uncertainty. It is also essential to discuss *confidence levels* based on concepts of probability and on distribution functions.

Before arriving at this, it is necessary to mention at least the notation commonly used to indicate the global uncertainty to associate with a measurement. This will be justified later on in several places in the text.

To indicate uncertainty and the measurement, we write:

$$x \pm \Delta x \tag{1.13}$$

meaning that it is highly plausible that quantity x, measured in a certain experiment or in several experiments performed one after another or independently in combined experiments, belongs to the interval of values

$-\Delta x, +\Delta x$.

To exemplify: if the global uncertainty were only one, obtained simply by estimation of casual uncertainty, $s = 0.2cm$, then, going back to the measurements in the previous chapter, we would have:

$$l \pm \Delta l = (8.1 \pm 0.2)cm \tag{1.14}$$

Or we could simply express only the reading error (equal to instrument resolution), for each measurement made with the meter. For example:

$$l_i \pm \Delta l_i = (7.7 \pm 0.1)cm \tag{1.15}$$

We will return to this concept of the expression of expanded uncertainty. We underline here that in the previous equation 1.14, the associated uncertainty should be equal to the expanded uncertainty determined by the casual uncertainty and evaluated as described in discussions to follow.

Example: the problem of the reading of zero with the polarimeter

We present here the case of the reading of zero or *offset*, when using the polarimeter, a common laboratory instrument.

The experiment is described in detail in appendix A.2. Here it is given simply as an example in which the declared accuracy of the instrument, and thus the reading error, as read on the scales, is much smaller than the systematic error introduced when taking measurements, not only due to the inadequacy of the instrument in its manufacture, but to the very principle of measurement. This is much less accurate than the scale appears to indicate.

Let us consider a normal polarimeter used in chemistry, biology and physics. In figure 1.5 we see the picture obtained in the so-called *equal-shadow* position in the eyepiece of the instrument. In this position the

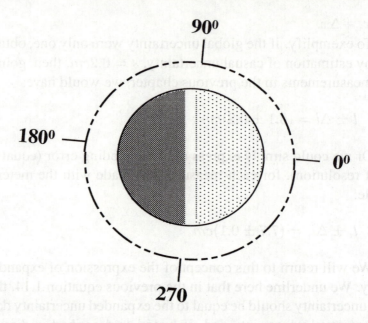

Figure 1.5: View through the eyepiece of *equishadow* in a position close to 0 on the graduated scale.

two half-moons should have the same intensity. At least the manufacturer refers to this position as the 0 position of the instrument if, for example, there is some de-ionised water inside the sample holder (see description of the instruments and experiments in Appendix A).

We could argue, however, that an *illegitimate* error caused by the difficulty of accurately determining light intensity in the two half-moons is quite plausible. For the rest of the discussion we will suppose that we are able to neglect this error, which in any case will make a contribution.

Let us consider the reading scale. The vernier should give a nominal resolution for the global reading of the scale, equal to $0°, 1', 30"$. Therefore, by repeating the measurement several times, the value of $0°, 0', 0"$ should be at the centre of the distribution, and the greater the number of

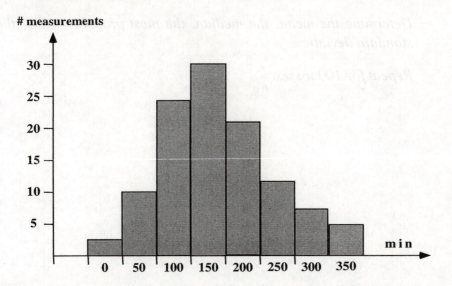

Figure 1.6: Distribution of values of the angles around the equishadow position. The scale is $1/60$ of a degree.

measurements the closer it will be to the centre. The data are given in figure 1.6. It is clear that the distribution is not symmetric and its centre is not a $0°, 0', 0"$. Since the number of measurements is large (more than a hundred), we can state that there is a systematic error which does not make deviations symmetric around the mean. This indicates that the scale of the instrument is not centred around $0°, 0', 0"$, and the position of equi-intensity does not correspond to it (figure 1.5).

Exercise: *tossing of dice*

Take two dice, with faces numbered from l to 6; toss them 20 times.

Write down the number of times you obtain a number corresponding to a number in the table then draw a histogram.

Determine the mean, the median, the most probable value, the standard deviation.

Repeat for 100 tosses.

2   Distribution of data and errors

2.1   Probability and distributions

So far we have seen how to calculate standard deviation starting from experimental data. In the laboratory, experimentalists need to go further. Experimental procedures very often consist of formulating an hypothesis, making the measurements and evaluating whether they satisfy the hypothesis. Standard deviation as discussed so far would be useful only in evaluating the accuracy of measurements. But this would not be of much use if it were not possible to compare this value with the values given by an hypothesis. We can use standard deviation to verify, for example, the hypothesis which led us to make the measurements. The problem is quite similar to what we discussed before. We want to see how well a measurement can be made, in the idealisation that a true value of a measurement exists.

For the best evaluation of how close we are to the true value, we can use the standard deviation of the parent distribution rather than that of the experimental distribution. We should then first of all see which parent distribution best approximates the experimental one. We will discuss this later on by making use of the *maximum likelihood* concept chapter 4.2) between two distributions. Evaluation of maximum likelihood is performed quantitatively by making use of the parameters that fully characterise the distributions, like the medians, the standard deviations, etc. First we shall classify the possible parent distributions most commonly used in practice and then we shall go on to discuss how to use them in testing hypotheses.

2.2 Distribution functions

In the following we will introduce the functions that analytically describe parent distributions. These functions are quite similar to the ones in common use in statistics and probability theory. To justify them, an introductory discussion of the concepts of probability theory is necessary.

In most textbooks, casual uncertainty is introduced after a more or less long introduction dealing with probability theory. A rigorous discussion should start from first principles and follow up with adequate mathematical reasoning. In this case we wish to take the opposite approach and lead students along a path of inductive reasoning.

Following common sense, *probability* gives us the *plausibility* of an *event*. We can consider this sufficient to describe the concept and use this definition from here on. It is clear that we can ascribe a higher or lower probability to an event and to a certain error, or to the result of a measurement, depending on how plausible it is considered.

For the time being, let us indicate probability as an analytical function of a variable which is the event ($\wp(x) \equiv$ probability of event).

To exemplify:

Example: toss of a coin, once or more than once.

On tossing a coin, we know that $\wp(x) \equiv \wp(tails) = 1/2$ and $\wp(x) \equiv \wp(heads) = 1/2$. We also know that $\wp(tails) + \wp(heads) = 1/2 + 1/2 = 1 \equiv P_{MAX}$, this will be the maximum probability. We also say that $\wp(tails) \equiv \wp(heads) = 50\%$ therefore $P_{MAX} = 100\%$. For simplicity's sake we assume here that the coin cannot come to rest on its edge. This means that the probability that neither of these two events takes place will be zero. We then define as 100% probability the certainty that one of the two possibilities will take place.

In order to describe the plausibility of the result of the tossing of two coins, two variables are necessary. One is x for the event, the other is n

for the total number of coins. Similarly, we could have 1 coin and make $n = 2$ tosses. Here we will treat these two cases as equivalent, assuming that they are independent. We assume that the two coins are independent in the same toss (i.e., they don't interfere with each other while falling), and we consider two consecutive tosses (i.e., the direction of the second toss is not influenced by where the first coin landed) as independent. The probability of the single event, without any a priori choice, *heads* or *tails*, is then written as:

$$\wp(x, n) = (\frac{1}{2})^n = (\frac{1}{2})^2 \equiv 25\% \tag{2.1}$$

and the total probability:

$$P_{MAX} = 100\% = \tag{2.2}$$
$$= \wp(tail, 1) + \wp(head, 1) + \wp(tail, 2) + \wp(head, 2)$$

Let us now choose either heads or tails. The probability of head can be indicated as *success*, ($\zeta = 0.5$) and correspondingly of tail *failure* ($\iota = 0.5$). Since there are no other possibilities, $\zeta = 1 - \iota$. Thus in this case the probabilities are:

$$\wp(x, n, \zeta) = \wp(x, n, \iota) = 50\% \tag{2.3}$$

However it is not always simply $\zeta = \iota$. If, for example, the second coin has a defect favouring the outcome of tails about 30% more than the outcome of heads, in total $\wp(x, n, \iota) \equiv \wp(tail, 2, 2/3)$, while for the defective coin would be $\wp(x, n, \iota) \equiv \wp(tail, 1, 1/3)$.

In order to calculate total probability P_{MAX}, we must take into account that $n > 1$ coin and x can be different from 1. Or in n tosses, x are heads (without taking into account which coin is involved), which means x successes are required. For example, if we require that in one toss of the two coins only one of them is head, this is sufficient for a success.

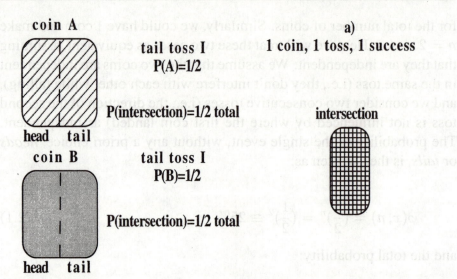

coin A

tail toss I
P(A)=1/2

a)
1 coin, 1 toss, 1 success

P(intersection)=1/2 total

intersection

head   tail

coin B

tail toss I
P(B)=1/2

P(intersection)=1/2 total

head   tail

Figure 2.1: Probabilities for different tosses and successes. $P(A)$, $P(B)$ and $P(C)$ indicate schematically a $\wp(x, n, \zeta)$. Case of any coin: the probability of a success is the intersection of the ensemble with itself.

Let us consider two cases. In one we make two tosses and obtain two heads. In this case the success is written as $\wp(x = 2 \; heads, 2, \frac{1}{2})$. In the other case, we make 2 tosses of 2 coins and we require that only 1 is heads. Then the success is written as $\wp(x = 1 \; head, 4, \frac{1}{2})$. It is clear that of the two cases, the second is *less probable*. In fact, it is $\wp(2, 2, \frac{1}{2}) = \frac{1}{4}$, $\wp(1, 4, \frac{1}{2}) = \frac{1}{16}$.

In reality, different definitions and concepts of probability lead us to consider the two cases above as different and not necessarily independent. Many texts on probability are to be found in the references. Let us now go on to discuss distribution functions.

Probability theory in more general cases will be discussed again in later chapters.

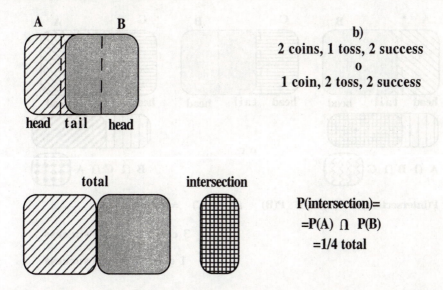

Figure 2.2: Probability for different tosses and successes. The $P(A)$, $P(B)$ and $P(C)$ indicate schematically a $\wp(x, n, \zeta)$. Case of two coins: the probability of two successes is the intersection of the two ensembles.

2.2.1   The Binomial distribution function

Let us now suppose that success is always represented by tails. We can calculate the probability of each event $\wp(x, n, \zeta)$ by referring to the concept of ensemble. We can see that a coin as a ensemble in the plane, like a rectangle divided into two equal parts representing heads and tails, or rather the probability of their appearing. In figures 2.1, 2.2, 2.3 are represented various cases, summarised in table 2.1 and table 2.2 which are both equivalent to the second one depending on whether or not we consider the coins or the tosses as different events.

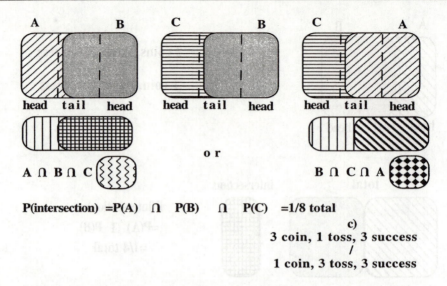

Figure 2.3: Probability of different tosses and successes. $P(A)$, $P(B)$ and $P(C)$ indicate schematically a $\wp(x, n, \zeta)$. Case of three coins: the probability of three successes is the intersection of one of the two intersections of the three ensembles.

In the following we will refer only to one of the two.

In figures 2.1, 2.2, 2.3 are represented three examples. $P(A)$, $P(B)$ e $P(C)$ They indicate schematically $\wp(x, n, \zeta)$.

It is to be kept in mind that the ensembles are obtained as a resulting probability, and not a total one. The intersection of the two ensembles is equal to the compositions of the probability depending on the product. The union of the two ensembles equals the composition depending on the sum.

Figure 2.1 (case a)) gives the overlapping between the two halves of the ensembles and the ensemble itself. Figure 2.2 (case b)) gives the overlapping between the halves of the two ensembles with respect to the total, i.e. to the two ensembles (the fact that there are physically two coins or two tosses brings to the same result). This intersection is equal to 1/4.

Similarly, in figure 2.3 (case c)), the intersection of the three ensembles is 1/8 of the total.

coins	tosses n	successes x	$\wp(x, n, \zeta)$
1	1	1	$\left(\frac{1}{2}\right)^1$
1	2	2	$\left(\frac{1}{2}\right)^2$
1	3	3	$\left(\frac{1}{2}\right)^3$
1	2	1	$2\left(\frac{1}{2}\right)^2$
1	3	1	$3\left(\frac{1}{2}\right)^3$

Table 2.1: Values of successes for the corresponding tosses of a coin.

Should the two not intersect and we wish to know what their composition is, we must keep in mind that the two ensembles can be unified only by intersecting, each one with the empty ensemble only. Therefore their composition is null. We can evaluate only total probability. Since each part is equal to 1/4, their union is equal to 1/2.

By iterating the process described in table 2.1, we obtain for the general formula as in table 2.3:

$$\wp(x, n, \zeta) = \frac{n!}{x!(n-x)!}(\zeta)^x(\iota)^{n-x} \tag{2.4}$$

more concisely:

$$\wp(x, n, \zeta) = \left(\begin{array}{c} n \\ x \end{array} \right) (\zeta)^x (\iota)^{n-x} \tag{2.5}$$

The $\wp(x, n, \zeta)$ given by equation 2.4 or by equation 2.5, is called *Binomial distribution function*.

The origin of the term gives an insight into its meaning. We will see below why it is so named.

Summing the function of all possible tries, n, one can combine the probability of all successes over all tries:

$$\sum_{x=0}^{n} \wp(x, n, \zeta) = \sum_{x=1}^{n} \left(\frac{n!}{x!(n-x)!} \zeta^x \iota^{n-x} \right) \tag{2.6}$$

This formula is quite familiar. In fact, for $n = 2$:

$$\sum_{x=0}^{2} \left(\frac{2!}{x!(2-x)!} (\zeta)^x (\iota)^{2-x} \right) = 2\zeta\iota + \zeta^2 + \iota^2 = (\zeta + \iota)^2 \tag{2.7}$$

one obtains the formula of the Binomial.

coins n	tosses	successes x	$\wp(x, n, \zeta)$
2	1	2	$(\frac{1}{2})^2$
3	1	3	$(\frac{1}{2})^3$
2	1	1	$2(\frac{1}{2})^2$
3	1	1	$3(\frac{1}{2})^3$

Table 2.2: Values of the successes with one toss and more than one coin.

Therefore equation 2.6 can also be written as:

coins n	tosses n	successes x	$\wp(x, n, \zeta)$
1	1	1	$\frac{1!}{1!(1-1)!}\left(\frac{1}{2}\right)^1\left(\frac{1}{2}\right)^{1-1}$
1	2	1	$\frac{2!}{1!(2-1)!}\left(\frac{1}{2}\right)^1\left(\frac{1}{2}\right)^{2-1}$
1	3	1	$\frac{3!}{1!(3-1)!}\left(\frac{1}{2}\right)^1\left(\frac{1}{2}\right)^{3-1}$
1	2	2	$\frac{2!}{2!(2-2)!}\left(\frac{1}{2}\right)^2\left(\frac{1}{2}\right)^{2-2}$
1	3	2	$\frac{3!}{2!(3-2)!}\left(\frac{1}{2}\right)^2\left(\frac{1}{2}\right)^{3-2}$
...	...	...	...
...	...	...	...
n	n	x	$\frac{n!}{x!(n-x)!}\left(\frac{1}{2}\right)^x\left(\frac{1}{2}\right)^{n-x}$

Table 2.3: Development of the Binomial expression for several coins and tosses. The first group refers to a coin, with the corresponding lines for different number of tosses and successes, up to the $n - th$ group of tosses with n coins.

$$\sum_{x=0}^{n} \wp(x, n, \zeta) = \sum_{x=1}^{n}\left(\frac{n!}{x!(n-x)!}(\zeta^x)(\iota)^{n-x}\right) = (\zeta + \iota)^n \quad (2.8)$$

if n are all possible tries, then equation 2.8 gives the total probability value, in fact:

$$\sum_{x=0}^{n} \wp(x, n, \zeta) = (\zeta + \iota)^n = (1)^n = 1 \qquad (2.9)$$

Exercise: *probability for one side of a die*

Throw the die 5 times with sides from 1 to 6.

Determine the probability of obtaining the number 6 twice.

Repeat for 100 tosses.

Exercise: *toss of coins*

Make 100 tosses with 10 coins.

Write in a table the number of times tail is obtained.

Construct the corresponding histogram.

Write the probability and verify the function of the Binomial distribution.

Repeat for heads.

We have thus obtained the function of probability which, as can be seen from the above exercise, describes the parent distribution which corresponds to an experimental distribution represented in figure 2.4.

One therefore says that *function of distribution* is more appropriate than *distribution of probability*. In the case of the exercise, we have a parent distribution, represented in figure 2.5 which is typical of the function of Binomial probability.

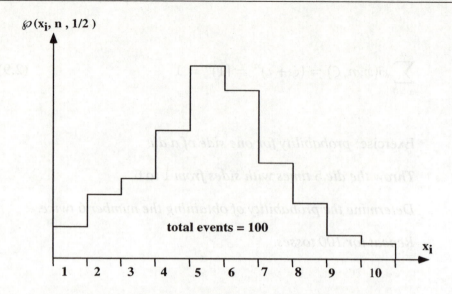

Figure 2.4: Experimental Binomial distribution of the exercise.

We shall see below that by changing the numerical value of the parameters, the shape of the curve of the function also changes, becoming more characteristic and distinguishable from the other distribution functions which will be introduced below.

With the data from the exercise with the 10 coins and 100 tosses, the product $\zeta \cdot n$ is approximately the arithmetical mean, in this case $0.5 \cdot 10$. We can demonstrate that this always holds true for a parent distribution and we can write:

$$\mu = n\zeta \tag{2.10}$$

In any case it is intuitive that on 10 coins about 50% are tails.

It must be kept in mind that in the previous exercise the number of coins, 10, should not be confused with the number of tries, 100, nor with the number of events. We must distinguish between the number of all possible values and the number n of the times the experiment was repeated. It is now the time to introduce an additional variable, such as an index in function $\wp$, now $\wp(x_i, n, \zeta)$ (figure 2.5), namely, the x_i are n in number, but they may vary between 1 and 10. However, for simplicity's sake we will omit the index in the following equations; this is also the most commonly used notation. Always to be borne in mind

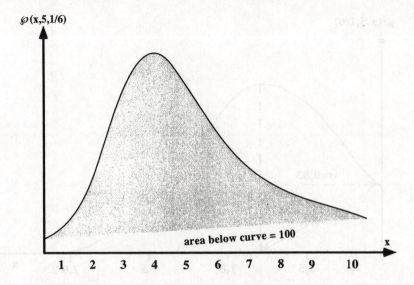

$\wp$(x,5,1/6)

area below curve = 100

x

1 2 3 4 5 6 7 8 9 10

Figure 2.5: The Binomial distribution parent to the experimental distribution in the example of the coins, with $\zeta = 1/2$.

is that x varies in the interval of possible values and with each event the $i-th$ we obtain a different value of x_i.

From the last exercise we can deduce the other distribution parameter, namely standard deviation. In the experimental distribution this equals about $s = 1.58$, as can be seen in figure 2.4. The same value can be obtained from formula:

$$\sigma = \sqrt{n\zeta(1-\zeta)} \cong s = \sqrt{10 \cdot \zeta \cdot \iota} = \sqrt{10 \cdot 0.5 \cdot 0.5} \qquad (2.11)$$

This expression is general for the standard deviation of a probability function of the Binomial distribution or more simply of the parent distribution. This expression is frequently used.

Let us now examine an example that leads to an approximation of 2.11.

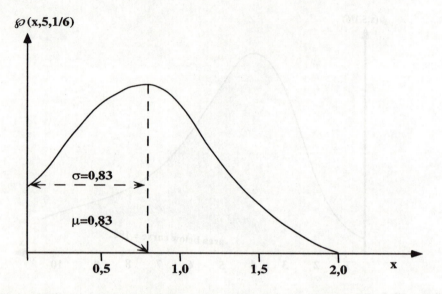

Figure 2.6: Binomial distribution parent to the experimental distribution in the example of the die with $\zeta = 1/6$ and $n = 5$.

Exercise: *the distribution parameters of the toss of the die*

Let us go back to the exercise of the die tossed 5 times with 2 successes of side 6. Now we shall discuss the parent distribution and not the experimental one. In fact, we do not have the experimental distribution of the toss of the die and of successes. We calculate the characteristic parameters according to the previous formula. We then have

$$\mu \cong n\zeta = 5 \cdot \frac{1}{6} = 0.83 \qquad (2.12)$$

and

$$\sigma \cong \sqrt{n\zeta(1-\zeta)} = \sqrt{5 \cdot \frac{1}{6}(1 - \frac{1}{6})} \cong \qquad (2.13)$$

$$\cong \sqrt{(0.83)(0.83)} = 0.83 \equiv \mu \qquad (2.14)$$

In this case we have $\zeta = 1/6$, so μ and σ coincide, contrary to the previous case of the coins, in which $\zeta = 1/2$. In figure 2.6, note that the shape of the curve is strongly asymmetric.

If we now change another parameter of the curve, for example the number of trials from 5 to $n = 15$, we obtain the curve in figure 2.7

Here we can see that the shapes are strongly dependent on parameter values. It is to be noticed in particular that for $n = 15$, $\sigma \neq \mu$. The shapes of the curves are so specific that a new function, still obtained as an approximation of the Binomial distribution, is suitable in describing several of them.

Example: the case of uncertainty on the efficiency of detection for the counting experiments

Let us consider an experiment very often used as an example and described in detail in the appendix A.3.

With a source of Sr^{90} one or more detectors made out of plastic scintillators are irradiated with electrons. The signal created by the passage of the particle is collected in a photomultiplier and sent to an electronic unit that gives a count proportional to the number of particles, which

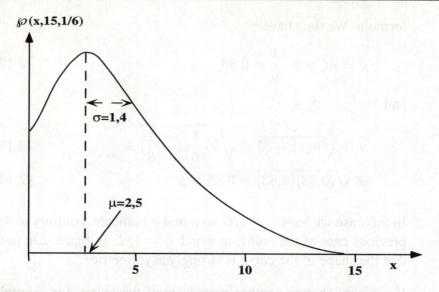

Figure 2.7: Binomial distribution parent to the experimental distribution in the example of the die with $\zeta = 1/6$ e $n = 15$.

have actually reached the detector. In measuring radioactive decay, one has to determine the number of particles produced by the decay (for example 100). Thus the mean number of particles emitted in a given time interval, $\overline{x} \cong \mu$ and the uncertainty to associate with this number is $s = \sqrt{\overline{x}}$. The result is usually indicated as *particles emitted* $= \overline{x} \pm s$ $\equiv \overline{x} \pm \Delta x = (100 \pm 10)$ (we will see in the following section why the uncertainty of counts is their square root).

It is necessary to determine the *efficiency* of the detector by taking into account that not all particles are in fact captured and may escape detection. To know how many particles are effectively emitted in the process of decay in the fixed interval, we must evaluate the efficiency.

Let us consider a detector with a counter of reference that indicates how much the first one counts with respect to itself. We consider the reference counter as the one that gives the number of particles emitted by the source. Let us consider the ratio of *number of particles counted/number of particles emitted* (the number indicated by #), which we call:

$$efficiency \equiv \varepsilon = \frac{\#counted}{\#emitted} \tag{2.15}$$

Therefore ε is always less than 1. For example, suppose that ϵ is equal to 20%, which means :

$$\varepsilon = \frac{20counted}{100emitted} = 0.2 \tag{2.16}$$

The uncertainty on ε is obtained using the parameters of the Binomial distribution function. This is a typical case of application of Binomial distribution because the value of parameter ζ is:

$$\zeta = 1 - \iota$$
$$\zeta = \iota$$

In fact, there is the same probability of a particle being detected as not being detected; indicating by $\zeta \equiv$ *particle detected* (*success*) and $\iota \equiv$ *particle not detected* (*failure*):

$$s \cong \sqrt{n\zeta(1-\zeta)} = \sqrt{100\frac{1}{2}(1-\frac{1}{2})} = \sqrt{25} = 5 \tag{2.17}$$

which means

$$s_\varepsilon \equiv \Delta\varepsilon = \frac{s}{100} = \frac{5}{100} = 5\% \tag{2.18}$$

which means

$$\varepsilon \pm \Delta\varepsilon = (20 \pm 5)\% \tag{2.19}$$

This is a general case for the distribution function. We will see in what follows that other distributions have a different value for the parameter of probability.

2.2.2 The Poisson distribution function

In the previous example, we examined the case of μ which decreases with respect to n and when n increases.

Since μ is equal to $n\zeta$, we can also have the same case in which, for two different experiments, n increases, μ is larger and ζ is smaller. If the number of trials n becomes large and μ remains small or decreases, then $\zeta \ll 1$. The function obtained from equation 2.5 [14] is called the *Poisson distribution function*, with the approximation

$$\zeta \ll 1 \tag{2.20}$$

we obtain

$$\wp(x, n, \zeta) \cong \wp(x, \mu) = \frac{\mu^x}{x!}e^{-\mu} \tag{2.21}$$

and for the standard deviation:

$$\sigma = \sqrt{n\zeta(1 - \zeta)} = \sqrt{\mu(1 - \zeta)} \cong \sqrt{\mu} \tag{2.22}$$

This equation is widely *ab*used. Let us give an example that will be discussed further on the following pages.

Example: the standard deviation in distributions from nuclear decays.

In nuclear decays the number of events can be very small. It may therefore happen that $\zeta \ll 1$, even with n fairly large [15]. The

---

[14]Full proof is inessential here because the student can see the consequences immediately. On the other hand, full proof can be found in many textbooks and is also included in the references.

[15]This statement holds only under certain conditions. For example, in the observation times which we will use later this will be a relevant issue. In this case they are not essential.

number of times a decay is detected has an uncertainty which can be expressed as the standard deviation of the parent distribution of the decays, which, assuming the approximation of the Poisson equation, is $\sqrt{\mu}$. To be noticed is the fact that the number of detections is given by the integral of the Poisson distribution function.

Example: The case of uncertainty on the height of bins in the counting experiment

In the counting experiment we can take the counts for a certain interval of time, for example, $1s$. We obtain a histogram similar to the one indicated in figure 2.8.

In this histogram, the bins indicate the number of times a certain count is repeated at different time intervals, all intervals being independent. Each measurement is therefore a count by itself. Together, the measurements form a Poisson distribution.

However, the single bins of the histogram may be considered as a Poisson distribution of mean value μ equal to the height of the bin, which are the *counts* which are in fact the *number* (#) *of times* [16] that such count (in the abscissa) has been detected (in the ordinate). Each count may be considered as a Poisson distribution in which $\zeta << 1$. This is also the only value of the distribution that corresponds to the arithmetic mean, namely the mean value of the parent distribution in which σ is the $\sqrt{\mu}$. Therefore, uncertainty on the height of the bin for one count is exactly $\sqrt{\mu}$, or better $\sqrt{\bar{x}}$.

If one has several counts, the bin can still be considered a Poisson

---

[16]Very often this quantity is called *frequency* and the one which is later on introduced as frequency is often called *relative frequency*. This terminology confuses the student. It is in fact not appropriate given its similarity to the physical quantity *frequency*.

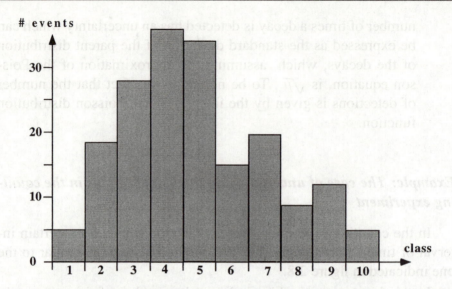

Figure 2.8: Histogram of the counts of a plastic scintillator irradiated from a source of Sr^{90} in consecutive time intervals equal to $1s$, for a total of 200 counts. The classes of counts are reported in the abscissa, the number of times the counts were detected in the ordinate.

distribution for which it is still true that $\zeta \ll 1$. Uncertainty [17] is no longer the square root of the count, but the square root of the number of times the count has been detected i.e. $\sqrt{\#times}$. In this second case the mean value of the distribution is the height of the bin, which is also the event itself, *the success* being that that number counts has been detected. In terms of probability, it is similar to having, in the first case, the number of times one has obtained heads with a coin; or, in the second case, the number of times one has obtained heads with several coins. Following the same notation, this corresponds to having

---

[17] A further demonstration of this fact can be found in the combination of the uncertainties 3.1.

$\wp(x, 1, \zeta)$ or $\wp(1, m, \zeta)$.

In any case, one always obtains a Binomial distribution, which with approximation $\zeta \ll 1$, is a Poisson one, which is reasonable if one maintains the fact that the counts are rare, that is, there are few in the single bins or namely that the single bins *are not very high*.

Once we have verified that for the single bin of the histogram the condition $\zeta \ll 1$ has been satisfied, the expression of uncertainty as equal to $\sqrt{\#times}$ is legitimate.

Example: comparison between uncertainties for the Binomial $\leftrightarrow$ Poisson distribution in the counting experiment

The other relevant quantity in the counting experiment is the *frequency* or *relative frequency* of the counts themselves, defined as

$$\varphi_i = \frac{counts_i}{counts_{total}} \tag{2.23}$$

We recall here that by $counts_i$ we mean the sum of each <u>number</u> (in the previous example indicated as $\#times$) of particles counted in each time interval. By $counts_{total}$, we mean the sum of the $counts_i$ in all the time intervals in which a measurement has been taken in the whole sample, for example 200, if the measurements were repeated 200 times for $1s$. This number is not the total number of counted particles. The number of counted particles is another important parameter in the distribution and plays an essential role in the discussion on the test of hypotheses in the following chapters. If the distribution was a parent and continuous one, would correspond to the area below the curves.

As we did for the counts, we can plot the corresponding value of the frequencies φ as a function of the *value of the count* or *class* of count in a graph having a single variable, and not in a histogram.

As we have already seen, uncertainty on the height of the , counts in one bin, is the square root of the height of the bin itself for the Poisson distribution. For the frequency, on the other hand, uncertainty is $\Delta\varphi$ and is given by the σ of the Binomial distribution. In fact, the single values of the frequencies have a Binomial distribution and not the ensemble of measurements, now expressed in frequencies, but still distributed according to the Poisson distribution. What we have just stated can be clarified by considering the range of values of the frequency itself. This goes from 0 to 1, namely from $0/200$ to $200/200$, in this example or, as it is often described as a percentage, from 0% to 100%. If a frequency is for example $\varphi = 40/200$, which is to say $\varphi = 20\%$, in the plot, the corresponding count $i = 4$, can also be interpreted as the probability of success (the measurement of the *the $i - th$* count) evaluated over the total number of counts.

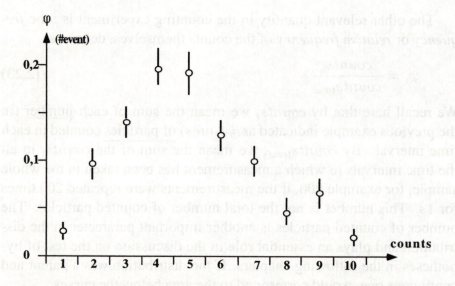

Figure 2.9: Plot of the frequency of counts of the same Poisson distribution as in the previous figure. The uncertainties bars derived from equation $\sigma = \sqrt{n\varsigma(1 - \varsigma)}$ are indicated.

Therefore, the uncertainty indicated in the plot in figure 2.9 is given by :

$$\Delta\varphi = \sqrt{n\zeta(1 - \zeta)} \tag{2.24}$$

For example

$$\Delta\varphi = \sqrt{200 \cdot 0.2(1 - 0.2)} \cong 1.4\% \cong 1\% \tag{2.25}$$

therefore

$$(\varphi \pm \Delta\varphi) = (20 \pm 1)\% \tag{2.26}$$

It is now worthwhile to add a comment on the distribution of frequencies. In fact, this distribution, like the one in figure 2.8, is a Poisson distribution (see figure 2.9). In the latter the values of ϕ are often indicated by the $\wp(counts_i, 200, \zeta)$ which presupposes identifying the concept of *theoretical* frequency with that of probability. This is a peculiar and debatable approach, known as *frequenist* [18] which in fact is based on a definition of probability as the quantity to which the theoretical frequencies (not experimental ones!) tend by increasing n. We can see here that the identification of $\wp$ with ϕ, shows at least two discrepancies. First of all, the fact that the experimental $\phi_{experimental}$ should tend to the $\phi_{theoretical}$ if $\zeta \ll 1$ and $n \to \infty$ and this fact is in contradiction with the fact that $\phi_{theoretical}$ should also tend to $\wp$ if $\zeta \ll 1$ and $n \to \infty$, because the distinction between theoretical frequencies and probability would be insignificant and therefore would make the frequentist definition meaningless.

In conclusion, the approximations and the definitions are highly debatable.

2.2.3 The Gauss distribution function

With an approximation different from the one made for Binomial distribution to define the Poisson distribution, we obtain another noteworthy distribution, represented by the curve in figure 2.10.

---

[18]This is commonly used, but it is not exact.

When

$$\zeta \cong 1 \qquad (2.27)$$

e

$$n \to \infty \qquad (2.28)$$

then

$$\mu = n\zeta \to \infty \qquad (2.29)$$

the probability distribution function is written as:

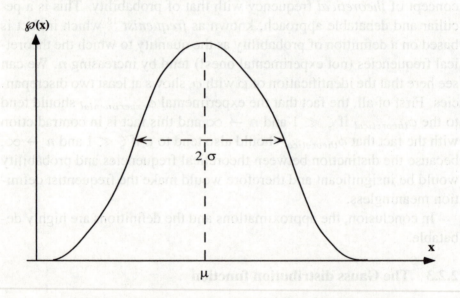

Figure 2.10: Parent distribution of the Gauss probability function.

$$\wp(x, n, \zeta) \equiv \wp(x, \sigma, \mu) = e^{-\frac{1}{2}(\frac{x-\mu}{\sigma})^2} \qquad (2.30)$$

In this equation the dependence of $\wp$ on μ and σ is explicit. In it, we do not have a simple general relation like equation 2.22, which connects μ and σ, so we therefore keep the variables separated.

More precisely the latter two are the very parameters of the distribution, with the variable x (or x_i) which can assume possible values in the defined interval. In the more general case x can assume all values between $-\infty$ and $+\infty$. In practice, for an experimental distribution, all possible values will be in a much more restricted interval, but a generalisation is useful in obtaining the general parent distribution function to describe the characteristic quantities of the curve frequently used in practice.

Coming back to the more general definition, we must remember that $\wp(x)$ is a probability for the single event. The total probability P_{TOT} that is obtained, in the hypothesis of non correlation between variables [19] (which is to say all possible independent events; for each event there is only one x and there cannot be any others different from that one, nor can any value be deduced from others), is the sum of the single $\wp(x)$. Since x is a variable of the parent distribution, having stated that it can assume all possible values between $-\infty$ and $+\infty$, we can also say that x is a continuous variable. To obtain total probability P_{TOT}, we sum up the continuous variable, which is to say the integral between $-\infty$ and $+\infty$.

$$P_{TOT}(x, \sigma, \mu) \equiv P(x, \sigma, \mu) = \int_{-\infty}^{+\infty} e^{-\frac{1}{2}\left(\frac{x-\mu}{\sigma}\right)^2} dx \qquad (2.31)$$

From the curve in figure 2.10, we can see that the shape is the same as the one in figure 1.3 obtained by the extrapolation of the number of measurements repeated and independent $n \to \infty$, all tending to the true value μ. The curve of figure 2.10 thus has the same shape as the parent

---

[19]This is tantamount to saying: in the hypothesis that the ensembles of figures 2.1, 2.2, 2.3 do not overlap, therefore the probabilities can be composed, that is, only summed and not multiplied.

one in an experimental distribution for $n \to \infty$, which is the same condition for the Gauss probability distribution function. *We can then identify the two.* Namely, we can *identify the value of a measurement with the value of the variable of the probability function.* We shall return to this concept at a later time.

Let us now examine some relevant consequences of this result. First of all it is quite useful to notice that, since at ∞ $x \to \mu$, one can indicate in the equation 2.31, the P_{TOT}, as:

$$P_{TOT}(x, \sigma, \mu) \equiv P(x, \sigma, \mu) \equiv P(\sigma) = \int_{-\infty}^{+\infty} e^{-\frac{1}{2}(\frac{x-\mu}{\sigma})^2} dx$$

(2.32)

in order to solve this integral, we must recall that:

$$F(t) = \int_{-\infty}^{+\infty} e^{-t^2} dt = \sqrt{\pi}$$

(2.33)

Namely, in this case:

$$F(t) = \int_{-\infty}^{+\infty} e^{-kt^2} dt = \sqrt{\frac{\pi}{k}}$$

(2.34)

Therefore, since

$$k = \frac{1}{2\sigma^2}$$

(2.35)

Then, on solving the integral [20] equation 2.32, we obtain:

$$P(\sigma) = \int_{-\infty}^{+\infty} e^{-\frac{1}{2}(\frac{x-\mu}{\sigma})^2} dx = \sqrt{2\pi}\sigma$$

(2.36)

---

[20]This is a *noteworthy* integral; it can be found in different textbooks, as well as in the references.

Taking into account that the single values of $\wp(x)$ represent the probability of the single event and therefore that P_{TOT} is the integral probability, P_{TOT} must necessarily be 100 %, that is, $P_{TOT} = 1$. It is for this reason that the Gauss distribution function is usually defined, instead of using equation 2.30, by

$$\wp(x, \sigma, \mu) = \frac{1}{\sqrt{2\pi}\sigma}e^{-\frac{1}{2}(\frac{x-\mu}{\sigma})^2} \tag{2.37}$$

which is the same, multiplied by the coefficient [21] $\frac{1}{\sqrt{2\pi}\sigma}$ which is characteristic of each distribution, obtained by the inverse of the result of the integral in equation 2.36.

In this way, we again define P_{TOT}, no longer $P_{TOT}(x, \mu, \sigma)$, but the *normalised* one $\Phi(x, \mu, \sigma)$:

$$\Phi(x, \mu, \sigma) = \frac{1}{\sqrt{2\pi}\sigma}P(\sigma) = \int_{-\infty}^{+\infty} \frac{1}{\sqrt{2\pi}\sigma}e^{-\frac{1}{2}(\frac{x-\mu}{\sigma})^2} dx = 1 \tag{2.38}$$

Another way to see this relation is, briefly:

$$x \to \mu$$
$$\sigma \to 0$$
$$\Phi(x, \mu, \sigma) \to 1$$

The first statement is justified by the definition of standard deviation σ, equation 1.9, since this one is the very difference between measured values and true values, which means between x_i and μ, then $\lim_{n \to \infty} \sigma = 0$ and $\sigma \to 0$ when $n \to \infty$.

It is best to warn students that in the following it is necessary to distinguish clearly $\wp$ from Φ.

---

[21] Often ostentatiously called the *normalisation coefficient or normalisation factor*. It is simply a factor that ensures that the area is numerically equal to 1.

Some noteworthy properties of the Gauss-type function

We shall now discuss several peculiar properties of the Gauss distribution function. Some of these are typical of the analytical expression, and are thus simply mathematical properties, but very useful in practical applications.

First, let us notice that for the first derivative of the function

$$f(x) = \frac{1}{\sqrt{2\pi}\sigma} e^{-\frac{(x-\mu)^2}{2\sigma^2}} \tag{2.39}$$

which is of the type:

$$f(x) = be^{-ax^2} \tag{2.40}$$

whose first derivative is :

$$\frac{df}{dx} = -2abe^{-ax^2} \tag{2.41}$$

we can write:

$$\frac{df}{dx} = -\frac{2(x-\mu)}{\sigma^2} \cdot \frac{1}{2\sigma\sqrt{2\pi}} e^{-\frac{(x-\mu)^2}{(2\sigma^2)}} = \tag{2.42}$$

$$= -\frac{(x-\mu)}{\sigma^3\sqrt{2\pi}} e^{-\frac{(x-\mu)^2}{(2\sigma^2)}}$$

From this equation, we can determine the flex point of the function, when the first derivative is zero, namely :

$$\frac{df}{dx} = 0 \tag{2.43}$$

when

$$x = \mu \tag{2.44}$$

which corresponds, in fact, to the maximum of the curve, since the third derivative is negative, as we will see later. This can also be seen by deriving for the determination of the flex points. We have for:

$$\frac{d}{dx}(\frac{df}{dx}) =$$

$$= (\frac{1}{\sigma\sqrt{2\pi}})\frac{d}{dx}(-\frac{x-\mu}{\sigma^2}e^{-\frac{(x-\mu)^2}{(2\sigma^2)}}) =$$

$$= (\frac{1}{\sigma\sqrt{2\pi}})((e^{-\frac{(x-\mu)^2}{(2\sigma^2)}}\frac{d}{dx}(-\frac{x-\mu}{\sigma^2})+(-\frac{x-\mu}{\sigma^2})\frac{d}{dx}(e^{-\frac{(x-\mu)^2}{(2\sigma^2)}})) =$$

$$= (\frac{1}{\sigma\sqrt{2\pi}})((e^{-\frac{(x-\mu)^2}{(2\sigma^2)}}(-\frac{1}{\sigma^2})+(-\frac{x-\mu}{\sigma^2})(-\frac{x-\mu}{\sigma^2})(e^{-\frac{(x-\mu)^2}{(2\sigma^2)}})) =$$

$$= (\frac{1}{\sigma\sqrt{2\pi}})(-\frac{1}{\sigma^2}+\frac{(x-\mu)^2}{\sigma^4})(e^{-\frac{(x-\mu)^2}{(2\sigma^2)}})$$

(2.45)

It is to be noted that this second derivative for $x = \mu$ is negative:

$$\frac{d}{dx}(\frac{df}{dx})_{x=\mu} =$$

$$= (\frac{1}{\sigma\sqrt{2\pi}})(-\frac{1}{\sigma^2} + frac(x-\mu)^2\sigma^4)(e^{-\frac{(\mu-\mu)^2}{(2\sigma^2)}}) =$$

$$= (\frac{1}{\sigma\sqrt{2\pi}})(-\frac{1}{\sigma^2}) < 0$$

(2.46)

This justifies what we said earlier, that for $x = \mu$ there is a maximum. Moreover, this second derivative is zero when

$$\frac{(x-\mu)^2}{\sigma^4} - \frac{1}{\sigma^2} = 0$$

(2.47)

namely when

$$x^2 - 2\mu x + \mu^2 - \sigma^2 = 0$$

(2.48)

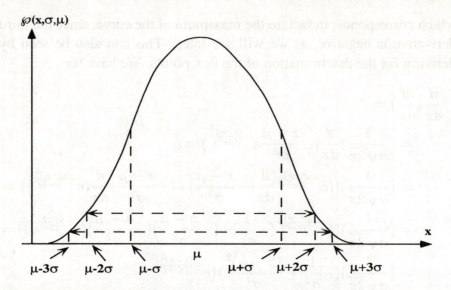

Figure 2.11: Parent distribution of the Gauss probability function. Values of the integrals are indicated in smaller intervals, corresponding to the abscissas $x = \mu + \sigma$, $x = \mu + 2\sigma$, $x = \mu + 3\sigma$.

which is verified for

$$x = \mu \pm \sigma \tag{2.49}$$

We therefore have two flex points for $x = \mu + \sigma$ and $x = \mu - \sigma$.

Let us now go to figure 2.11. In the plot we see the abscissas for $x = \mu + \sigma$, $x = \mu + 2\sigma$, $x = \mu + 3\sigma$ and $x = \mu - \sigma$, $x = \mu - 2\sigma$, $x = \mu - 3\sigma$.

The area of these intervals is widely used. To calculate it, it is necessary to solve the integral which, however, does not have a simple primitive function and thus requires a numerical integration and tabulating.

In the intervals indicated, the values are:

$$\Phi(x = \mu - \sigma, \ldots, x = \mu + \sigma) =$$

$$= \int_{x=\mu-\sigma}^{x=\mu+\sigma} \frac{1}{\sigma\sqrt{2\pi}} (e^{-\frac{(x-\mu)^2}{(2\sigma^2)}}) \cong 0.6827 \quad (2.50)$$

$$\Phi(x = \mu - 2\sigma, \ldots, x = \mu + 2\sigma) =$$

$$= \int_{x=\mu-2\sigma}^{x=\mu+2\sigma} \frac{1}{\sigma\sqrt{2\pi}} (e^{-\frac{(x-\mu)^2}{(2\sigma^2)}}) \cong 0.9545 \quad (2.51)$$

$$\Phi(x = \mu - 3\sigma, \ldots, x = \mu + 3\sigma) =$$

$$= \int_{x=\mu-3\sigma}^{x=\mu+3\sigma} \frac{1}{\sigma\sqrt{2\pi}} (e^{-\frac{(x-\mu)^2}{(2\sigma^2)}}) \cong 0.9973 \quad (2.52)$$

Therefore, in these intervals, it is about 68.3%, 95.5%, 99.7%, of the total.

It is to be noted that the various integrals can be expressed as percentages and the fact that they assume that specific numerical value is due to coefficient $\sigma\sqrt{2\pi}$. If the latter factor changed, the values would change.

Going back to figure 2.11, these values of the abscissas correspond, thanks to this very normalisation, to the different areas below the curve, indicated in the graph in figure 2.12. If the curve is not normalised at 100%, the correspondence between the values of the areas and the ones of the abscissas of figure 2.11 is not verified.

The 68% of the area is between $\pm 1\sigma$. It is an estimate of the width of the curve, as was to be expected, since the width of the parent distribution is the standard deviation which in fact indicates how much the values of x *deviate* from the mean value. This quantity must not be confused with the *full width at half maximum*, which is obtained from the

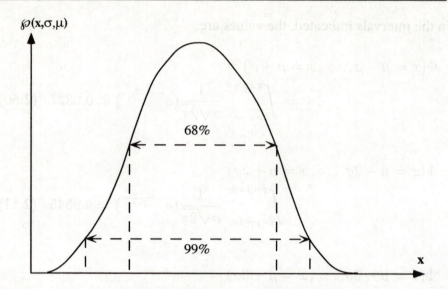

Figure 2.12: Values of the area below the curves of the Gauss probability function, in the intervals $(\mu - \sigma, \mu + \sigma)$, $(\mu - 3\sigma, \mu + 3\sigma)$.

plot by taking half of the most probable value ordinate and considering the interval between $-x$ and $+x$ when

$$\wp(x) = \frac{1}{2}\wp(x = \mu) =$$

$$= \frac{1}{2}\frac{1}{\sigma\sqrt{2\pi}} \neq \frac{1}{(\sqrt{e})}\frac{1}{\sigma\sqrt{2\pi}} \cong \frac{1}{1.65}\frac{1}{\sigma\sqrt{2\pi}} \qquad (2.53)$$

The integration intervals indicated above and the corresponding values of the abscissas will be used often in what follows.

The values of the ordinates, namely the values of $\wp(x; \mu, \sigma)$, are also noteworthy and help in practical applications. Let us recall first of all that for $x = \mu$, the function $\wp(x; \mu, \sigma)$ holds:

$$\wp(\mu; \mu, \sigma) = \frac{1}{\sigma\sqrt{2\pi}} \qquad (2.54)$$

Furthermore, for $x = \mu - \sigma$, the function $\wp(x = \mu - \sigma; \mu, \sigma)$ holds:

$$\wp(\mu - \sigma; \mu, \sigma) = \frac{1}{\sigma\sqrt{2\pi}} \frac{1}{\sqrt{e}} \tag{2.55}$$

i.e. :

$$\wp(\mu - \sigma; \mu, \sigma) \equiv$$
$$= \frac{1}{\sqrt{e}} \wp(\mu; \mu, \sigma) \cong 0.6065 \wp(\mu; \mu, \sigma) \tag{2.56}$$

and similarly for $\wp(\mu + \sigma; \mu, \sigma)$. Similarly:

$$\wp(\mu - 2\sigma; \mu, \sigma) \equiv \wp(\mu + 2\sigma; \mu, \sigma) =$$
$$= \frac{1}{\sigma\sqrt{2\pi}} \frac{1}{e^2} = \frac{1}{e^2} \wp(\mu; \mu, \sigma) \cong 0.1353 \wp(\mu; \mu, \sigma) \tag{2.57}$$

and

$$\wp(\mu - 3\sigma; \mu, \sigma) \equiv \wp(\mu + 3\sigma; \mu, \sigma) =$$
$$= \frac{1}{\sigma\sqrt{2\pi}} \frac{1}{e^{4.5}} = \frac{1}{e^{4.5}} \wp(\mu; \mu, \sigma) \cong 0.0111 \wp(\mu; \mu, \sigma) \tag{2.58}$$

Many times we also find a different indication of the values of the integrals in the above-mentioned intervals. They too will be quite useful later on. One usually indicates noteworthy values taken from tables, those for which the area corresponding to the interval are respectively:

$$area 90.0\% \qquad x \pm 1.645 \cdot \sigma \tag{2.59}$$
$$area 95.0\% \qquad x \pm 1.960 \cdot \sigma \tag{2.60}$$
$$area 99.0\% \qquad x \pm 2.576 \cdot \sigma \tag{2.61}$$
$$area 99.9\% \qquad x \pm 3.290 \cdot \sigma \tag{2.62}$$

The corresponding values are indicated in the plot of figure 2.13.

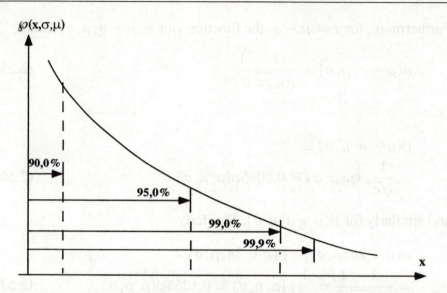

Figure 2.13: Illustration of the right tails of the areas of the Gauss probability function for different intervals of σ. The values of integrals are indicated in increasingly large intervals, in correspondence to the values of the areas of 90.0%, 95.0%, 99.0%, 99.9%.

Example: The case of the constant τ in the experiment on circuit RC

Let us now examine an example of how the exponential function, used also in the case of the Gauss function, is often used and how its parameters, described in the above section, have an immediate physical interpretation.

We shall see in what follows an experiment on nuclear decay processes, which follow an exponential law. This is exactly the same as one part of the Gauss function or like one of the two halves of a bell. We see here an easy application of the noteworthy values of the exponential function.

We saw earlier that on multiplying by a factor equal to $\frac{1}{\sqrt{e}}$ the maximum value of the curve, we obtain the value of the corresponding ordi-

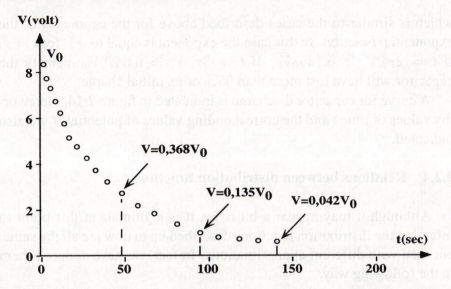

Figure 2.14: Representation of the exponential law for a capacitor discharge. The values of V are indicated as percentages with respect to V_0 for the corresponding values of τ.

nate for $x = \mu - \sigma$ then we can deduce σ.

In Appendix A.4 there is the description of a simple experiment on a RC circuit with a resistor and a capacitor in series or parallel. Here we study a specific example. We wish to study the capacitor discharge. This process follows an exponential law:

$$V = V_0 e^{-\frac{t}{\tau}} \tag{2.63}$$

where V_0 and V are the initial value of the potential and the ones in subsequent time intervals t and $\tau = RC$ is the constant of characteristic discharge time, which depends on the capacity and resistance of the circuit. It is therefore the interval of time after which at $t = \tau$:

$$V = V_0 e^{-\frac{\tau}{\tau}} = \frac{1}{e} V_0 \cong 0.3679 V_0 \tag{2.64}$$

which is similar to the cases described above for the exponents of the exponential function. In this case the exponent is equal to -1, for $t = \tau$. If $t = 2\tau$, $V \cong 0.1353V_0$. If $t = 3\tau$, $V \cong 0.0479V_0$: namely the capacitor will have lost more than 95% of its initial charge.

A curve for capacitor discharge is indicated in figure 2.14. Noteworthy values of time t and the corresponding values of potential V are also indicated.

2.2.4 Relations between distribution functions

Although it may appear a bit risky, it is legitimate at this point to infer that the distribution functions described up to now are all the same, but seen with different approximations. In fact, we have deduced them in the following way.

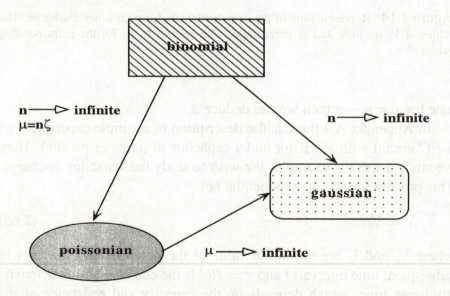

Figure 2.15: Schematic representation of relations between the Gauss, Poisson and Binomial distribution functions.

From the Binomial distribution, approximating the relation

$$\wp(x, n, \zeta) = \frac{n!}{x!(n-x)!}(\zeta)^x (\iota)^{n-x} \tag{2.65}$$

with approximation

$$\zeta \ll 1 \tag{2.66}$$

we obtain the Poisson distribution function and from the Binomial distribution we obtain the Gauss distribution function, under the condition

$$\zeta \cong 1 \qquad n \to \infty \qquad \mu \to \infty \tag{2.67}$$

Therefore, as in the diagram in figure 2.15, we can think of the same distribution applied to different situations, for example for an ensemble of experimental data.

Since we can choose a priori to have a large number of measurements, we have

$$n \to \infty \tag{2.68}$$

and also with values much different from zero, namely with a large mean value,

$$\mu \to \infty \tag{2.69}$$

The same natural processes, as we shall see, may however be quite rare, as in the example of radioactive decay. Therefore neither the number of counts with small mean value μ, nor the number of measurements n, would verify the approximation for the Gauss distribution function, while it would on the contrary be applicable, for example, in the case of the Poisson distribution function.

These comments must not induce students to confuse different approximations with different distribution parameters of the same type.

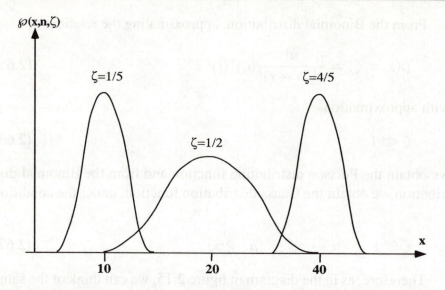

Figure 2.16: Comparison between different shapes of curves of the Binomial distribution function, for the same values of n and different values of ζ.

A distribution is not to be identified with a specific function if we do not know the values of its parameters and these are not evaluated within the experimental situation at hand, we cannot infer the shape or the function. For example, in figure 2.16 are indicated different Binomial distributions, with the same n and different ζ and vice versa, with same ζ and different n, in figure 2.17.

The curves may have the same shape as a Gauss distribution function, but not respect the approximations for that reason.

Another example in figure 2.18 stresses again the importance of parameter evaluation.

The three distribution function curves have different values of ζ and different values of n, with $\mu=n\zeta$, small (equal to 2.5), which does not vary greatly with the varying of n, which, instead, becomes quite big (much bigger than μ).

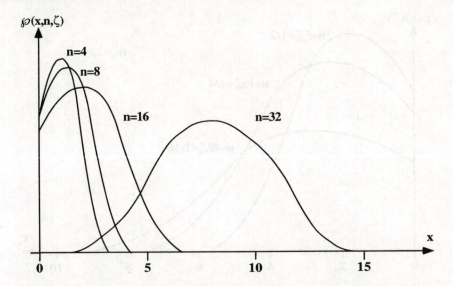

Figure 2.17: Comparison between different shapes of the Binomial distribution function curves, for different values of n and the same values of ζ.

The same curves can be compared to the ones in the previous figures, or to a Gauss distribution function (figure 2.19).

Though they are similar in shape, the difference between the two curves is substantial. For the Gauss curve we can see that the negative values are allowed, whilst for the Poisson one these are not allowed. If we think of these as distribution functions of experimental data, for example, the Poisson one as a count distribution, we understand that we cannot have negative values of the counts, so the continuous curve cannot be a Poisson curve.

It is to be pointed out that very often these comparisons are made only for the sake of justifying the abuse of the parameters of the Gauss distribution function, for all experimental situations. But it is not legitimate to state that all experimental distributions tend to, or may, approximate a Gauss one. The discussion just concluded explains why this procedure does not have a solid base.

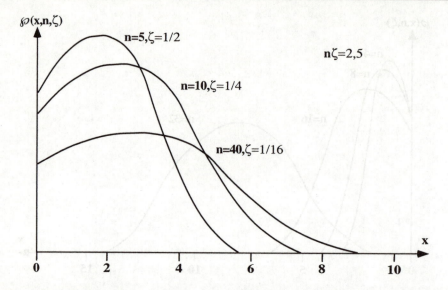

Figure 2.18: Comparison between different shapes of the curves of the Poisson distribution function, for different values of n and different values of ζ.

In the following we give an example that illustrates the application of the Poisson and Gauss distribution functions and their characteristic parameters.

Example: the case of the Poisson $\rightarrow$ Gauss distribution in the counting experiment

Let us consider the counting experiment. Let us suppose that we perform two series of measurements, keeping the source in the same position and the rest of the apparatus in the same conditions for the collection of counting data in consecutive time intervals equal to $1s$, as described in the previous example.

We now collect a series of data in much longer time intervals: $45s$.

Following what we said before, for the approximation to one Gauss

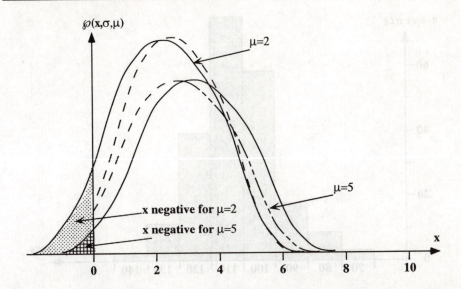

Figure 2.19: Comparison between different shapes of the curves of the Poisson and Gauss distribution functions, for the same ζ and the same n, in the two cases for $\mu = 2$ and $\mu = 5$, that is, by varying either ζ or n. The dashed-line curves indicate the Poisson function, the solid-line curves indicate the Gauss function.

distribution function, the counts are definitely *less rare*: namely $\zeta \cong 1$. Furthermore we naturally expect to collect a much greater number of counts: namely, $\mu \to \infty$.

The counts obtained for a total of 200 measurements are indicated in figure 2.20, similar to figure 2.8.

We shall now go on to verify rigorously that the parent distribution to this experimental distribution is described by a Gauss distribution function. It is clear that the shape of the curves is clearly more symmetric and quite different from the one obtained for time intervals of $1 s$.

For the arithmetic mean of the counting classes we can write that:

$$\overline{x}_{45s} \gg \overline{x}_{1s} \tag{2.70}$$

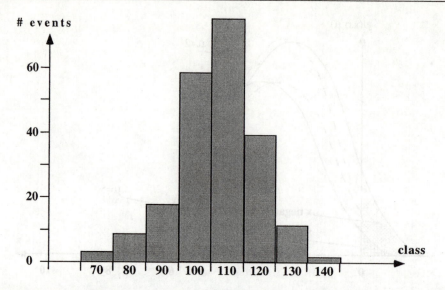

Figure 2.20: Histogram of the counts by a plastic scintillator irradiated by a source of Sr^{90} in different time intervals equal to $45s$, for a total of 200 counts. In the abscissa are shown the *classes* of counts, in the ordinate the number of times these counts were measured.

In the same manner it is clear that the total number of counts are :

$$n_{45s} \gg n_{1s} \tag{2.71}$$

In conclusion, we have seen a case in which the Binomial distribution function, in its Poisson approximation, tends towards the Gauss distribution on increasing the width of the measurement interval, while maintaining the same apparatus and the same procedure.

We have therefore seen that the distribution function is not a property of the physical process, but rather a property of the measurement process.

2.2.5 The constant distribution function and the reading error

We introduce here a particularly simple distribution function, widely used in practice.

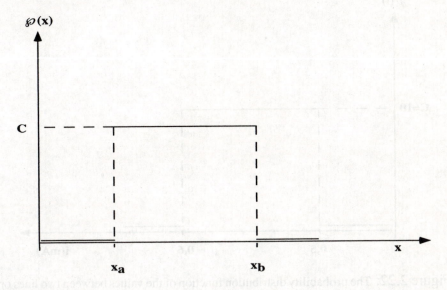

Figure 2.21: Curve of a constant distribution function for each value of x in the interval between x_a and x_b.

Let us consider a function of a variable x, which can assume, for example, the values of a series of measurements. Let us suppose that we obtain the same value $y_c = f(x_i) = constant \equiv C \ \forall \ x_i$, in a certain interval between x_a and x_b, as in figure 2.21. In general, the variable may be continuous.

The function may also be a probability distribution function for which all values are *equiprobable*.

Let us suppose that the interval of values is between two lines on a scale. For example, it may be the interval between $0.5mA$ and $0.6mA$ in a tester with a *maximum value* of $10mA$.

Let us consider the uncertainty for the reading error of the instrument: with this scale, by evaluating only the interdistance between two marks, it is $0.1mA$.

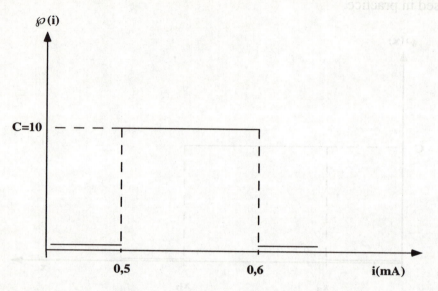

Figure 2.22: The probability distribution function of the values between two lines on a tester. The values of the extremities of the interval, $x_a = 0.5mA$ and $x_b = 0.6mA$, corresponding to the values of the lines on the scales are indicated.

With this distribution, let us now verify rigorously if the uncertainty value associated with it is correct. It is legitimate to ascribe a certain probability value to each measurable value. In particular, it is legitimate in this case to ascribe a probability to each measurable value included between the two marks. This value will in fact be equi-probable.

Not knowing which value is the true one between $0.5mA$ and $0.6mA$, we can assume that all values have the same probability.

Therefore we can represent the values between $0.5mA$ and $0.6mA$ in the abscissa and their probability in the ordinate, represented in figure 2.22.

The probability distribution function corresponds to the possible values between the two lines, only in that interval. The other values cannot be taken into account, because they are readable with the other lines on the scale, those to the left and to the right. The function must therefore be such that $f(x) = C$ for $x \geq x_a$ and $x \leq x_b$ and $f(x) = 0$ for all the other values of x.

Let us examine here the parameters of the probability distribution. The mean value μ corresponds with the median, and the $f(\mu)$ with the most probable value, in any case constant. We now have to determine the σ of the distribution.

Before doing that, we must note that we can ensure that the area under the curve, which is a rectangle, is equal to 1, so as to maintain the analogy with the probability distribution functions.

The constant C of the ordinates in the interval x_a, x_b must be such that the area of the rectangle (simply $base \cdot height$) is equal to 1. Therefore :

$$(x_b - x_a) \cdot C = 1 \tag{2.72}$$

Therefore one has:

$$C = \frac{1}{x_b - x_a} \tag{2.73}$$

We can demonstrate simply that σ of this distribution is

$$\sigma = \frac{x_b - x_a}{\sqrt{12}} \tag{2.74}$$

In this example, let us call δ:

$$\delta = x_b - x_a = 0.6mA - 0.5mA = 0.1mA \tag{2.75}$$

the difference is equal to the width of the interval of values, that is, to the distance between two consecutive lines.

It is therefore legitimate to assume that since all values are equiprobable in the interval between the lines, the true value is the mean value μ, in the center of the interval.

The σ of the distribution is :

$$\frac{0.1}{\sqrt{12}} \tag{2.76}$$

Let us now return to the reading error and the resolution. This (see chapter 1) is often considered the interval or the semi-interval between the lines. We can see here that the σ is smaller than half of the interval semi-interval:

$$\frac{1}{\sqrt{12}} \cong 0.288675 < 0.500000 \equiv \frac{1}{2} \tag{2.77}$$

It constitutes a different estimate of the resolution of the instrument, namely, of the reading error. In fact, σ is the standard deviation of the distribution, in analogy to what we wrote for the Gauss one, and this will be discussed again later on in the discussion on confidence intervals.

One has:

$$\sigma \cong \frac{0.1}{\sqrt{12}} \cong 0.03 \tag{2.78}$$

Therefore σ is the evaluation of the width of the equiprobability curves. In this meaning it is a better estimate of uncertainty than δ, since it is the best estimate of deviations from the mean value, of all possible measurements. Therefore a measured value between two lines can be written as:

$$(0.56 \pm 0.03)mA \tag{2.79}$$

having approximated $\frac{0.11}{\sqrt{12}} \cong 0.03$ and having truncated the approximation at the central value of the x at the second digit after the decimal point.

Here one could write $(0.56 \pm 0.03)mA$, but this would have been in contradiction with the estimated accuracy, as calculated above. It would also need an accuracy in the calculation of the mean, better than the one of the measuring device which is by itself, the best possible available for the measurement. This is the reason of writing $(0.56 \pm 0.03)mA$ instead of $(0.56 \pm 0.03)mA$.

Here a word about significant digits is in order. It would be unreasonable to estimate even more digits than the second digit after the decimal point. This would not give any further information because in any case the second digit after the decimal point is uncertain.

In conclusion, we have identified a method for a more rigorous determination of the resolution of the instrument equipped with lines for the identification of uncertainty due to the reading error.

We therefore state that if the two lines are readable enough and the reading needle is particularly thin with respect to the inderdistance between the lines, and the whole scale is reasonably visible to the naked eye, an additional digit can be added to the uncertainty of the measurement. It is clear that if the manufacturer thought that the instrument could indicate values with greater accuracy this would be indicated on the scale, with adequate specifications. The value of equation 2.74 has to be considered a lower limit of reading uncertainty than an instrument equipped with a scale at the conditions given above.

Example: the case of underestimating the reading error in the experiments on photoresistance

The problem of the reading error, can be better illustrated by the following example, which shows that often it is not important to try to appreciate too small intervals, even when they are easily seen on the scale.

We shall describe in detail an experiment in the study of the response of a resistor made out of semiconductor material, silicon, with the varying of incident light.

The resistor can be illuminated with a lamp in an optical system aligned and focalised with two suitable polarising filters, placed between the lamp and the resistor which allow variation of the intensity of the incident light.

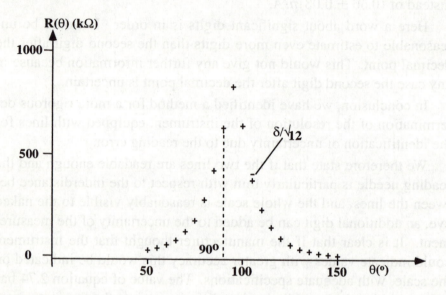

Figure 2.23: Curve of resistance values of a *Si* photoresistor with variation of the rotation angle of the polariser, which varies incident light intensity. The maximum is clearly shifted by 10° with respect to the expected central position of 90°. The figure also shows the presumed reading uncertainty.

The filters are equipped with rotating graduated scales. Therefore the measurement of intensity is indirectly given by the angles of filter rotation. We can measure variation in resistance value at the two ends of the photoresistor by varying angle ϑ of rotation of one of the two filters.

The scale of one filter has 36 lines in the interval between 0° and 180°, with an index of the same size as the lines. Therefore, the shortest distance between the lines is 5°.

By applying the described criterion, we would have a resolution that can be identified with uncertainty due to the reading error equal to:

$$\Delta\vartheta = \frac{5°}{\sqrt{12}} \cong 1°27' \tag{2.80}$$

Actually, on making the measurements, we find that the maximum of light intensity corresponds to the value $\vartheta \cong 80°$, instead of 90°, as is reasonably to be expected in a symmetrical system.

This indicates that there is a systematic error due to the off-axes of the system which is not easily visible. A similar curve (not indicated in figure 2.23), taken in different light conditions, shows a shift of different value in correspondence to the maximum. By using the value of equation 2.80 as reading uncertainty we do not find a real correspondence between the accuracy of the measurement of the apparatus and the accuracy expected by the experimenter. Therefore, in these experimental conditions, the reading error estimated analytically from equation 2.80, is too small with respect to the uncertainty equal to the width of the interval between two lines.

We shall see that with rigorous analytical verification of a law that describes the experimental curve, underestimation of the reading error seriously compromises the success of accuracy tests, thus invalidating them.

3 Processing of data and uncertainties

3.1 The combination of uncertainties

The final result to be reached is not always the outcome of a direct experimental measurement. The quantity is often derived from other quantities. The uncertainty to be associated with it will therefore be derived from others as well. From here on, we shall introduce a formal way which has direct and immediate application in obtaining the value of the uncertainty in a rigorous way, by combining several uncertainties together.

Students often consider derived relations as formulas to be applied by rote in the operations to be performed with the uncertainty value and do not appreciate the fact that each time they must be derived for each application. It is to be kept in mind that the relations derived below are simple factorizations of a single and unique rigorous way of combining uncertainties, which is the subject of this section.

It is therefore better not to rely on formulas learned by heart, but simply try to understand the method [22].

3.1.1 Rigorous calculation for the estimate of the combined uncertainty

Let us consider a measurement of volume derived from three measurements of length. Let us suppose we have obtained these measurements through different repeated ones for the three dimensions.

We have:

the value of width l_i

the value of height h_i

---

[22]This is misleadingly called *propagation* as if the value of the uncertainty were identified with the idea of the propagation of a virus.

the value of depth p_i

The volume is the quantity measured, to which we wish to associate a certain uncertainty:

$$x_i = x_i(l_i, h_i, p_i) \equiv V_i(l_i, h_i, p_i) \tag{3.1}$$

or in general

$$f_i = f_i(u_i, v_i, \dots) \equiv x_i(u_i, v_i, \dots) \tag{3.2}$$

having obtained by direct measurements only $u_i, v_i, \dots$ and having calculated the arithmetic mean:

$$\bar{x} = \bar{x}(\bar{u}, \bar{v}, \dots) \tag{3.3}$$

It is natural that the uncertainty of the derived quantity is a generic function of the uncertainties of the other quantities measured:

$$\sigma_{\bar{x}} = \sigma_{\bar{x}}(\sigma_{\bar{u}}, \sigma_{\bar{v}}, \dots) \tag{3.4}$$

as well as:

$$\sigma_x = \sigma_x(\sigma_u, \sigma_v, \dots) \tag{3.5}$$

since

$$\sigma_x^2 = \lim_{n \to \infty} \sum_{i=1}^{n} \frac{1}{n}(x_i - \bar{x})^2 \tag{3.6}$$

In order to have an estimate of uncertainty, we can consider the variance of 3.5 and 3.6, as a function which can be approximated with a Taylor series. We have to develop the sum of the Taylor series around the deviations $d_i = (x_i - \bar{x})$. This operation can be performed for a function of one variable, as well as for a function of more than one variable.

Given the function $f(x)$, its development *around* point x_0 is obtained from:

$$f(x) \cong f(x_0) + (\frac{\partial f}{\partial x})_{x=x_0}(x - x_0) +$$

$$+ \frac{1}{2}(\frac{\partial^2 f}{\partial x^2})_{x=x_0}(x - x_0)^2 + \cdots$$

$$(3.7)$$

which for two variables can be written as:

$$f(x, y) \cong x_0 + y_0 + (\frac{\partial f}{\partial x})_{x=x_0, y=y_0}(x - x_0) +$$

$$+ (\frac{\partial f}{\partial y})_{y=y_0, x=x_0}(y - y_0) + \frac{1}{2} \cdots$$

$$(3.8)$$

Let us now develop the function x, around itself, namely around $\bar{x}$, neglecting $\bar{u}$ and $\bar{v}$:

$$x_i \cong \bar{x} + \cdots$$

$$(3.9)$$

or, equivalently, we can approximate the deviations $d_i = (x_i - \bar{x})$:

$$(x_i - \bar{x}) \cong$$

$$\cong (u_i - \bar{u})(\frac{\partial x}{\partial u})_{v=\bar{v}} + (v_i - \bar{v})(\frac{\partial x}{\partial v})_{u=\bar{u}} + \cdots$$

$$(3.10)$$

Therefore substituting in 3.6:

$$\sigma_x^2 = \lim_{n \to \infty} \sum_{i=1}^{n} \frac{1}{n}(x_i - \bar{x})^2 \cong$$

$$\cong \lim_{\frac{1}{n} \to \infty} \sum_{i=1}^{n} ((u_i - \bar{u})^2 (\frac{\partial x}{\partial u})^2 + (v_i - \bar{v})^2 (\frac{\partial x}{\partial v})^2 + \cdots) \quad (3.11)$$

but

$$\lim_{n \to \infty} \left(\sum_{i=1}^{n} \frac{1}{n} (u_i - \bar{u})^2 \right) = \sigma_u^2 \qquad (3.12)$$

and

$$\lim_{n \to \infty} \left(\sum_{i=1}^{n} \frac{1}{n} (v_i - \bar{v})^2 \right) = \sigma_v^2 \qquad (3.13)$$

while the mixed term in the mixed product is equal to

$$\lim_{n \to \infty} \left(\sum_{i=1}^{n} \frac{1}{n} (u_i - \bar{u})(v_i - \bar{v}) \right) \equiv \sigma_{uv}^2 \qquad (3.14)$$

The latter equation defines a very precise quantity known as *covariance* σ_{uv}^2. This name comes from the similarity of the notation with variance and this wording comes from the matrix notation in which it is often included for the sake of brief notation. In the following we will introduce it in this way.

Covariance is of great importance in more complex applications than the combination of uncertainties. It is an essential and unique tool in evaluating the underestimation or the overestimation of uncertainties, if relations are not simple ones. It is a measurement of the degree of *correlations* between two or more uncertainties. Covariance gives an immediate evaluation of the interdependence between two or more uncertainties, that is, if an uncertainty is *in relation* or *correlated* with another. Its importance is clear, given the fact that without knowing it, we can either underestimate or overestimate uncertainty, thus invalidating the whole procedure of combination of uncertainties.

Covariance σ_{uv}^2 should not be confused with the product $\sigma_u \sigma_v$. Covariance is often defined a posteriori and evaluated only after a careful experimental analysis. Its importance is inversely proportional to our knowledge of the dependence of a certain quantity and its associated uncertainty on any other experimental measurement.

For the moment, for the sake of simplicity and in order to obtain a simple relation for the combination of uncertainties, we will make an approximation, which will also help us to arrive at a better understanding of what covariance is.

We will develop a logical procedure to show that covariance is a measurement of the degree of correlation between two uncertainties. The reasoning will be presented starting from the end. In fact, we start with the approximation that uncertainties are not correlated and then we show that this fact implies that covariance is equal to 0. Then we deduce the fact that covariance is not 0 if uncertainties are correlated.

It is to be noted that the deviations from equation $u_i - \overline{u}$ equation 3.14, are zero at ∞, by definition, as well as the deviations $v_i - \overline{v}$, but the product is not necessarily zero at∞. The first term of equation 3.14, can be written as:

$$\lim_{n \to \infty} \left(\sum_{i=1}^{n} \frac{1}{n} (u_i - \overline{u})(v_i - \overline{v}) \right) = \tag{3.15}$$

$$= \lim_{n \to \infty} \frac{1}{n} \left(\sum_{i=1}^{n} (u_i v_i - u_i \overline{v} + \overline{uv} - v_i \overline{u}) \right) \tag{3.16}$$

In order to visualize the case in which variables are correlated or not at ∞, we can, for simplicity's sake, consider two distributions centred at 0, namely with $\overline{u} = 0$ and $\overline{v} = 0$. Thus only the mixed product $u_i \cdot v_i$. is different from 0. There will be positive and negative values. Their algebric sum tends to 0 by increasing n. The student can verify this by evaluating the result of the sum with the increase in the number of pairs. This exercise can first be done for $u_1 \cdot v_1$, then for $u_1 \cdot v_1 + u_2 \cdot v_2$ and so on. The result of the sum is smaller and smaller. If also $\overline{u} \neq 0$ and $\overline{v} \neq 0$, the reasoning is quite similar.

If u_i and v_i, *are independent*, by going to the limit for $i = 1, \ldots, n$ with n at∞, we have as many u_i as v_i, so the cross products $u_i \cdot v_i$, taken with their algebraic sign, mutually go to zero, as well as their product for $u_i \cdot \overline{v}$ and $v_i \cdot \overline{u}$. Therefore the product to ∞, goes to zero but not the product $\bar{u} \cdot \bar{v}$. In general, equation 3.6, can be written as:

$$\sigma_x^2 \cong \sigma_u^2 \left(\frac{\partial x}{\partial u}\right)^2 + \sigma_v^2 \left(\frac{\partial x}{\partial v}\right)^2 + 2\sigma_{uv}^2 \left(\frac{\partial x}{\partial u}\right)\left(\frac{\partial x}{\partial v}\right) + \cdots \tag{3.17}$$

with

$$\sigma_{uv}^2 \to 0, n \to \infty \tag{3.18}$$

This means that covariance is zero at∞ if variables are not correlated. In the opposite case the cross products above may not go exactly to zero. If, for example, the v_i depend on u_i, even in a simple manner like:

$$v_i = 3.5u_i \tag{3.19}$$

the sum of the cross products may be different from zero, but in any case unknown. Thus it will be necessary in this case to evaluate covariance and use the complete equation 3.17, otherwise [23]:

$$\sigma_x^2 \cong \sigma_u^2 \left(\frac{\partial x}{\partial u}\right)^2 + \sigma_v^2 \left(\frac{\partial x}{\partial v}\right)^2 \tag{3.20}$$

Equation 3.20 is a guide in evaluating combined uncertainty in simple situations, in which errors can be approximated as uncorrelated.

In the following we shall show some interesting cases in which combined uncertainty can be obtained thanks to simple expressions that connect the variables together: the sum, the product, the power, and so on. It is clear that they are not formulas to be applied mnemonically and that they are deduced from the general expression 3.20 or, the more general one 3.17.

---

[23]It is very often written that in this case the assumption that the errors are *Gauss distributed* holds. This is very misleading terminology; one may ask *if the errors are Gauss-like they are not correlated, and if they are Binomial are they!?* One pretends that the variables are independent, not that they are Gauss distributed.

It is to be noted, on the contrary, that if covariance is zero, variables are not necessarily independent.

Exercise: *combination of the uncertainties in special expressions*

- approximated equation for the sum

$$x = au + bv \tag{3.21}$$

$$\frac{\partial x}{\partial u} = a \tag{3.22}$$

$$\frac{\partial x}{\partial v} = b$$

$$\sigma_x = \sqrt{a^2\sigma_u^2 + b^2\sigma_v^2 + 2ab\sigma_{uv}^2} \cong$$

$$\cong \sqrt{a^2\sigma_u^2 + b^2\sigma_v^2} \tag{3.23}$$

- approximated equation for the product

$$x = auv \tag{3.24}$$

$$\frac{\partial x}{\partial u} = av \tag{3.25}$$

$$\frac{\partial x}{\partial v} = au$$

$$\sigma_x = \sqrt{a^2v^2\sigma_u^2 + a^2u^2\sigma_v^2 + 2a^2uv\sigma_{uv}^2} \cong$$

$$\cong \sqrt{a^2v^2\sigma_u^2 + a^2u^2\sigma_v^2} \tag{3.26}$$

- approximated equation for the division

$$x = a\frac{u}{v} \tag{3.27}$$

$$\frac{\partial x}{\partial u} = \frac{a}{v} \tag{3.28}$$

$$\frac{\partial x}{\partial v} = -\frac{au}{v^2}$$

$$\sigma_x = \sqrt{\frac{a^2}{v^2}\sigma_u^2 + (-\frac{au}{v^2})^2\sigma_v^2 + 2(\frac{a}{v})(-\frac{au}{v^2})\sigma_{uv}^2} \cong$$

$$\cong \sqrt{\frac{a^2}{v^2}\sigma_u^2 + (-\frac{au}{v^2})^2\sigma_v^2} \tag{3.29}$$

Exercise: *combined uncertainty for special functions*

derive the approximated expressions of uncertainty for a quantity depending on one variable u and form parameters a and b, in the cases of

- power

$$x = au^b \tag{3.30}$$

- exponential

$$x = ae^{ub} \tag{3.31}$$

- logarithmic

$$x = a \ln ub \tag{3.32}$$

Example: The case of uncertainty on the length in the coaxial cable experiment

As an example of the practical application of the combination of uncertainties, we shall again consider the experiment on the measurement of the speed of light inside a coaxial cable

We have already discussed the case of the illegitimate error in the measurement of lengths in this experiment.

It is necessary to measure the length of a cable which is continually lengthened by adding pieces which should all have the same length. A simple carpenter's rule is used to measure the total length of the cable. For each measurement of length, the delay between two voltage signals is measured on an oscilloscope and, from the ratio between lengths and time intervals, we obtain the speed of light.

As we have already stated, it is reasonable to consider the reading error as the resolution on the scale of the rule and in this discussion we neglect illegitimate errors.

To perform the measurement of length, the rule needs to be moved a certain number of times, depending on the length of the cable. The final uncertainty for each measurement is obtained from the combination of uncertainty coming from the reading error (namely the resolution, however it may be defined), repeated as many times as the rule has been moved.

Let us suppose we have obtained the following measurement, for the length of a cable in three approximately equal pieces:

$$l = l_1 + l_2 + l_3 = 373.0 cm \tag{3.33}$$

The measurement actually should be performed on the whole length made out of the three pieces and not on the three single pieces (it is not relevant in this example of the combination of uncertainties, but it is important in reducing the *illegitimate errors*, discussed above to a minimum. By so doing, the number of times the meter has been repositioned is slightly higher than the number of added pieces. In this example, it has been positioned 4 times for 3 pieces because the whole rule is exactly one meter long while the cables are slightly longer. If, on the contrary, we had made the measurements separately for each piece of cable, the number of positionings would have been much higher than the number of measurements: $2 \cdot 3 = 6$, twice the number of pieces. Therefore the number of times the measurement is performed is inadequate in the latter case, or more precisely, the measuring instrument is inadequate, and its use has not been optimised. Since the meter is shorter than the length of the cable, the number of positionings which now correspond to our reading, must be reduced.

Let us go back to the numerical example, in the first case of unified measurements, we have:

$$\Delta l \cong \sqrt{(\Delta l_1)^2 + (\Delta l_2)^2 + (\Delta l_3)^2 + \cdots} \tag{3.34}$$

Before giving it a numerical value, it is useful to underline that it is possible in this example to see a kind of *correlation between the uncertainties*, as described in the previous section. This is because each measurement of length depends on the previous one in its positioning. Therefore we can see a correlation between the errors, not the reading errors, but the illegitimate ones. In this discussion we have assumed that these are negligible, but only, of course, in the situation in which we make every effort to avoid them while measuring.

For the moment it is sufficient to point this out about illegitimate errors. Let us go back to the combination of uncertainties, neglecting this correlation. For the first case we can write:

$$\Delta l = \sqrt{(0.1cm)^2 + (0.1cm)^2 + (0.1cm)^2} \cong \qquad (3.35)$$
$$\cong 0.173205cm \cong 0.2cm \qquad (3.36)$$

The discussion on the approximation of the digits will be continued in the following section.

By comparing these measurement with 6 positionings:

$$\Delta l \cong \sqrt{6 \cdot (0.1cm)^2} \cong 0.244949cm \cong 0.3cm \qquad (3.37)$$

As was to be expected, an increase in the number of positionings brings about an increase in uncertainty. Greater care taken in optimising measurements leads to a lower value for uncertainties, still keeping to a rigorous method in the combination of the single uncertainties.

In conclusion :

$$l \pm \Delta l = (373.0 \pm 0.2)cm \qquad (3.38)$$

3.2   The number of significant digits

Let us now discuss once again the previous example, which will be useful in addressing one of the most delicate issues concerning experimental measurements, *the number of significant digits*. This is a delicate issue because one often tries to give certain rules while actually it is common sense that can guide us in a measurement and its calculations. It is a good idea not to apply strict mnemonic rules but to evaluate the numerical value of the quantities with the relative uncertainties, before defining the number of significant digits [24].

Here we can see just how useful the combination of uncertainties is. Let us go back to the case of the combined uncertainty Δl. In this case we had:

$$\Delta l \cong 0.173205 cm \qquad (3.39)$$

and in the other

$$\Delta l \cong 0.244949 cm \qquad (3.40)$$

Neither of these numbers are significant in our calculations which deal with precise measurements of quite simple physical quantities obtained with simple instruments with a well-defined resolution to be combined. Since the resolution of the rule was equal to $\Delta l_i = 0.1 cm$, it makes no sense to combine values obtained with consecutive calculations with a number of digits more than one decimal place, because this would mean that we would have measured the quantity with an uncertainty lower than the resolution. This procedure is therefore meaningless. This, actually, gives the name to the number of digits. It does not matter whether they are before or after the decimal point; they will be called *significant digits* if they are actually numbers that can be measured and there will be as many as the estimate of their uncertainty will give significance to.

Students are often misled by calculations and operations with functions or simply by exponents. If we consider a case like the previous

---

[24] *When I was a student* ... I started laboratory just when pocket calculators started widespreading among students. For this reason teachers used to say that these instruments led to the use of many more digits than were necessary and lamented the demise of the slide rule and common sense. In reality, the old textbooks used when these teachers were students and before the advent of the pocket calculator and the computer contain statements on the use of significant digits just as misleading. One example: ... *The number of digits to be used in the laboratory is in any case a maximum one decimal place!*

one, on calculating the quantity $(0.1)^2$, to continue by approximating the result to the first digit after the decimal point, namely (0.1), does not make any sense for several reasons. First of all because it would overestimate the error and would not correspond to the reason why we introduced the calculation of the combination of uncertainties, which was to arrive at a reasonable estimate of the combined uncertainty. Going on with the calculation, in fact, we would obtain:

$$\Delta l \cong \sqrt{(0.1cm)^2 + (0.1cm)^2 + (0.1cm)^2} \cong$$
$$\cong \sqrt{(0.1)cm^2 + (0.1)cm^2 + (0.1)cm^2)} \cong$$
$$\cong 0.54772cm \cong 0.6cm \tag{3.41}$$

which is much larger than a reasonable uncertainty $(0.2cm)$ which we would obtain by repeating the measurement four times with a rule of $\Delta l_i = 0.1cm$ or six times with an uncertainty $\Delta l_i = 0.3cm$.

The other reason why this calculation is meaningless is that the numbers shown under the square root are lengths to the square. This means that the quantity has been obtained from an operation of the quantity multiplied by itself. Simply by applying what we learned earlier on the combination of uncertainty, this would mean combining quadratically the single uncertainties which are equal to the estimated resolutions. This gives :

$$\Delta(l_i^2) \equiv \Delta(l_i \cdot l_i) \cong \sqrt{(0.1)^2 + (0.1)^2} \tag{3.42}$$

We would then have to start the process again from the beginning. To interrupt it and state once again what we said earlier on, that what we wrote above in equation 3.42, is wrong, let us go back to the definition of the instrumental or reading error. We stated that this is the smallest appreciable uncertainty, and also the best estimate of it. If we ascribe an uncertainty to the best estimate, we recognise that in fact this is not the best estimate. We have thus demonstrated through absurdity the validity of the estimate in equation 3.36 and how wrong equation 3.42 was. In conclusion, it is meaningless to say that there is an uncertainty to associate also with the product of the two best estimates, or on its power.

What we have said so far should lead the student to a suitable evaluation of the significant digits exclusively on the basis of the uncertainties with which students know the measurements. However, they may find themselves in difficulty both because they do not perform the calculations with the measured quantities or not with physical quantities, and because their uncertainty is unknown. Any calculation can be based on the a priori knowledge of the uncertainty on this number. Supposing, for example, that we have to multiply two numbers in the tables given in textbooks, such as $a = 0.5468321$, with $b = 9.2$. We assume that all the digits up to the last decimal place indicated are known. At the most, the final result cannot give but the first significant digit after the decimal point: namely, $c = a \cdot b = 5.1$. If we were dealing with measured physical quantities, one with $\Delta a = 0.000001$ and the other one with $\Delta b = 0.1$, the combination of the uncertainties would give $\Delta c \cong 0.1$. Therefore, also in this case, we have only one significant digit after the decimal point, as stated previously.

3.3 The problem of correlation

Let us now go back to the problem of correlation. This chapter can be omitted in a first reading, but sooner or later it must be dealt with to have an idea of the meaning of covariance.

Let us consider a series of measurements u_i performed with an associated uncertainty equal to s_u. This, *not the measurements*, can be correlated with s_v, associated with very different measurements v_i.

Of course, this correlation must not be obvious to the experimentalist, otherwise it should be possible to intervene. For example, it would be possible to estimate it a posteriori, once it was discovered. In this case the uncertainty would then be separated into several parts by separating systematic uncertainties.

In practice, there are cases of more complex and derived measurements in which the correlation can be *inferred* and one has to ascribe a certain numerical value to it. This leads to a more or less complex operation on the standard deviations of the parent distribution, or with the usual approximation, of the experimental ones.

We can also have the opposite case in which, instead of writing the single uncertainties and the systematic effects separately, it is better to use the formalism shown here to describe the interdependence of uncertainties in a more succinct manner.

In our discussion on the combination of uncertainties, we stated that the deviations between the two series of measurements go to zero in the sum up to ∞. This is true if there is no dependence of one variable on the other. Otherwise it is necessary to introduce the covariance term as previously defined in equation 3.14:

$$\sigma_{uv}^2 \equiv \lim_{n \to \infty} \left(\sum_{i=1}^{n} \frac{1}{n} (u_i - \overline{u})(v_i - \overline{v}) \right) \tag{3.43}$$

To apply the complete expression of the combination of uncertainties, we approximate for a finite number of measurements:

$$\sigma_{uv}^2 \cong \left(\sum_{i=1}^{n} \frac{1}{n} (u_i - \overline{u})(v_i - \overline{v}) \right) \tag{3.44}$$

This term must be added to the expression which gives the uncertainty of the derived quantity $x(u, v)$. In practice, it is not so simple. The dependence between variables u and v must be *hidden*. If, on the contrary, there is a clear *relation* or *correlation* between the two quantities u and v, then the derived quantity x will be directly expressed or expressible as a function of one of the two u or v. What is more, the two variables can be considered independent *unless there are systematic errors*, namely, *if they have been measured as apparently uncorrelated but their measurements contain an uncertainty that is common to both of them.*

Having stated that, the student may find it a bit difficult to understand how and when to evaluate covariance in the expression of the uncertainty for the derived quantity. The evaluation and explicit expression of covariance is used more often as an expression for the experimental uncertainties on the degree of systematic interdependence of the measurements. The use which is common in statistical theory of the covariance quantity is much more general than the use introduced here. The term covariance is used just to express the dependence between more than one variable, not only uncertainties, evaluated separately to determine their interdependence quantitatively. The variables of qualitative entities, or simply of classes of measurements, very often do not have the exact quantitative characteristic of a physical measurement and/or an analytical function that describes it. Therefore in these case there are no simple analytical relations between them.

Here we consider only the case of the expression of covariance for uncertainties.

A practical example is the use of one tester in the experiment described in A.7 for the determination of the characteristic curve of a diode. The idea is to determine the curve $I = I(V)$, of current intensity as a function of the voltage applied to the diode. We can use two separate testers or the same tester in detecting voltage and the current, simply by changing scales. The tester may be poorly calibrated or simply show an *offset*, namely a value of zero on the scale that does not correspond to the true zero value of the quantity [25]. In this case the measurements and the reading errors of the instruments would show a correlation. Therefore, in this case covariance is not null.

Let us now see how covariance appears in the synthetic matrix form by introducing the *error matrix*. Very often the opposite situation is encountered, not of having to construct this matrix but of having to interpret it as a result of a computer programme. It is above all with this aim that we introduce it here.

---

[25]Of course all this must be unknown to the experimentalist a priori, otherwise he/she would change the tester; or it may be discovered only after taking the readings, which is often the case of students (and not only them).

3.3.1   The error matrix or matrix of uncertainties

Let us consider two series of independent measurements, which are distributed according to a Gauss parent distribution, each one centered around its mean value $\mu_u = \mu_v \equiv 0$ (figure 3.1):

$$\wp(u; \mu_u, \sigma_u) = \frac{1}{\sqrt{2\pi}\sigma_u} e^{-\frac{1}{2}\frac{u^2}{\sigma_u^2}} \tag{3.45}$$

and

$$\wp(v; \mu_v, \sigma_v) = \frac{1}{\sqrt{2\pi}\sigma_v} e^{-\frac{1}{2}\frac{v^2}{\sigma_v^2}} \tag{3.46}$$

Since the two distributions are independent (see the discussion on probabilities for independent events in the part devoted to Binomial distribution and also the one in the following chapter on maximum likelihood), their product is given by the combined probability:

$$\wp(u, v; \sigma_u, \sigma_v) = \frac{1}{2\pi\sigma_u\sigma_v} e^{-\frac{1}{2}(\frac{u^2}{\sigma_u^2} + \frac{v^2}{\sigma_v^2})} \tag{3.47}$$

which can be represented in a three-dimensional plot. In figure 3.2, in the ordinate we find the values of $\wp(u, v; \mu_u, \mu_v, \sigma_u, \sigma_v)$. In figure 3.3 we see the plane that intersects the surface at $\wp(u = 0, v = 0; \mu_u, \mu_v, \sigma_u, \sigma_v)/\sqrt{e}$. The intersection describes an ellypse.

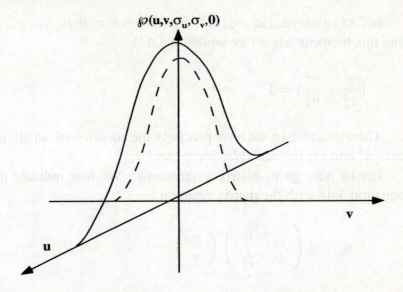

Figure 3.1: Curve of distribution of the two variables u and v, with mean value equal to zero.

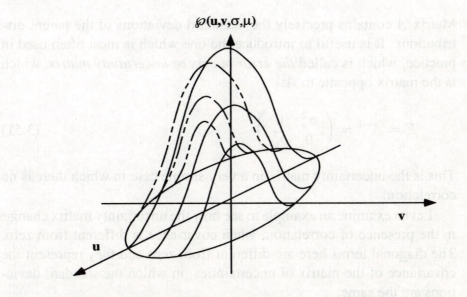

Figure 3.2: Convolution of the distribution curves of the two variables u and v.

In fact $\wp(u, v; \mu_u, \mu_v, \sigma_u, \sigma_v) = \wp(u = 0, v = 0; \mu_u, \mu_v, \sigma_u, \sigma_v)/\sqrt{e}$, and this happens when (see equation 3.47)

$$(\frac{u^2}{\sigma_u^2} + \frac{v^2}{\sigma_v^2}) = 1 \tag{3.48}$$

Therefore at $\pm 1\sigma$ we have precisely the equation of an ellypse, with axes u^2 and v^2; *the variables are σ_u and σ_v.*

Let us now go to matrix expressions. We first indicate the same equation 3.48 with the matrix notation.

$$\begin{pmatrix} u & v \end{pmatrix} \begin{pmatrix} \frac{1}{\sigma_u^2} & 0 \\ 0 & \frac{1}{\sigma_v^2} \end{pmatrix} \begin{pmatrix} u \\ v \end{pmatrix} = 1 \tag{3.49}$$

This in matrix symbols becomes:

$$V \cdot A \cdot V = 1 \tag{3.50}$$

Matrix A contains precisely the standard deviations of the parent distributions. It is useful to introduce the one which is most often used in practice, which is called *the error matrix* or *uncertainty matrix*, which is the matrix opposite to A:

$$E = A^{-1} = \begin{pmatrix} \sigma_u^2 & 0 \\ 0 & \sigma_v^2 \end{pmatrix} \tag{3.51}$$

This is the uncertainty matrix in a very simple case in which there is no correlation.

Let us examine an example to see how the uncertainty matrix changes in the presence of correlation, when covariance is different from zero. The diagonal terms here are different from zero and they represent the covariance of the matrix of uncertainties, in which the standard deviations are the same.

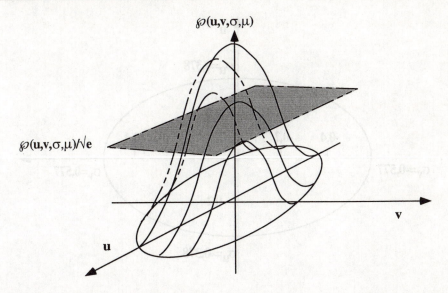

Figure 3.3: Plane intercepting $\wp(u = 0, v = 0; \mu_u, \mu_v, \sigma_u, \sigma_v)\, e^{-1/2}$ the convolution of the two distribution curves in u and v. The intersection describes an ellipse.

Example [26]: the uncertainties ellipse.

Let us suppose that σ_u and σ_v have very precise numerical values.

$$\sigma_u = 0.378$$
$$\sigma_v = 0.577$$

therefore

$$\frac{u^2}{\sigma_u^2} + \frac{v^2}{\sigma_v^2} = 1 \qquad (3.52)$$

---

[26] A similar example can be found in [15] taken from [18], where it has been used in a completely different environment with respect to the discussion here. We simply take the algebric formulae here.

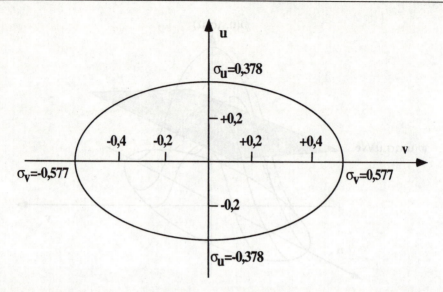

Figure 3.4: Example of an ellipse, with axes equal to $\sigma_u = 0.378$ and $\sigma_v = 0.577$, when there is no correlation between the variables, and therefore between standard deviations.

i.e.

$$\frac{u^2}{(0.378)^2} + \frac{v^2}{(0.577)^2} = 1 \tag{3.53}$$

which means

$$\frac{u^2}{(0.143)} + \frac{v^2}{(0.333)} = u^2(7) + v^2(3) = 1 \tag{3.54}$$

This is the equation of the ellipse indicated in figure 3.4.

As we said above, if there is no dependence between the two variables, even if hidden, there is no dependence between the standard deviations. For this reason, and only as an illustration of how to compare and represent covariance in the uncertainty matrix, in this example, we artificially introduce a dependence, or relation or *correlation* between two variables, but keeping in mind that we are interested in the correlation and how this is *transferred* to the standard deviations.

Let us consider the same distributions in u and v, now with dependence of v on u and vice versa. With respect to figure 3.1, they will appear rotated. In the same way the ellipse will appear so, as in figure 3.4.

Let us introduce an artifical rotation of the two axes of the ellipse.

Analytically, this is equivalent to changing the system of coordinates with a transformation, which is a rotation of the new coordinates u' and v'.

For example, depending on the rotation angle, the equation of the ellipse 3.54 may become:

$$6u'^2 + u'v' + v'^2 = 1 \qquad (3.55)$$

which in matrix form can be written:

$$V' \cdot M \cdot V' = 1 \qquad (3.56)$$

namely

$$\begin{pmatrix} u' & v' \end{pmatrix} \begin{pmatrix} 6 & 1 \\ 1 & 1 \end{pmatrix} \begin{pmatrix} u' \\ v' \end{pmatrix} = 1 \qquad (3.57)$$

In this case, we have to rewrite the uncertainty matrix. This is, as we said before, the inverse of M: $E = M^{-1}$.

Before doing that we must briefly recall the procedure for inverting any matrix A:

$$A = \begin{pmatrix} a_{11} & a_{12} \\ a_{21} & a_{22} \end{pmatrix} \qquad (3.58)$$

the corresponding inverted matrix $B = A^{-1}$,

$$B = \begin{pmatrix} b_{11} & b_{12} \\ b_{21} & b_{22} \end{pmatrix} = A^{-1} \tag{3.59}$$

is obtained by first calculating

$$det A = a_{11}a_{22} - a_{21}a_{12} \tag{3.60}$$

and then transposing and dividing by $det A$:

$$B = \begin{pmatrix} \frac{(-1)^{2+2}a_{22}}{det A} & \frac{(-1)^{2+1}a_{21}}{det A} \\ \frac{(-1)^{1+2}a_{12}}{det A} & \frac{(-1)^{1+1}a_{11}}{det A} \end{pmatrix} = A^{-1} \tag{3.61}$$

Therefore, since in this example

$$det M = 6 - 1 = 5 \tag{3.62}$$

the uncertainty matrix E is :

$$E = M^{-1} = \frac{1}{5} \begin{pmatrix} 1 & -1 \\ -1 & 6 \end{pmatrix} \tag{3.63}$$

Explicitly, therefore the standard deviations are:

$$\sigma_{u'}^2 = \frac{1}{5} = 0.200$$

$$\sigma_{v'}^2 = \frac{6}{5} = 1.200$$

therefore the values of the axes of the ellypse in the new reference system have changed:

$$\sigma_{u'} = 0.447$$

$$\sigma_{v'} = 1.096$$

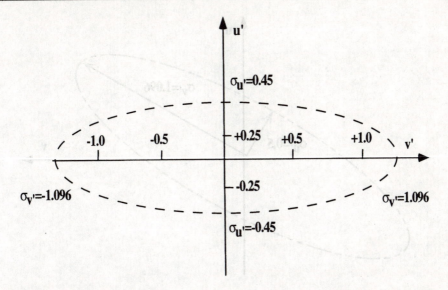

Figure 3.5: Axes of the ellipse rotated by an angle ϑ. The new values of the axes are indicated.

This is indicated in figures 3.5 and 3.6 in which the ellipse is simply rotated by an angle ϑ with respect to axes u and v.

The term

$$\frac{-1}{5} \equiv \sigma^2_{u'v'} \qquad (3.64)$$

is precisely the *covariance*.

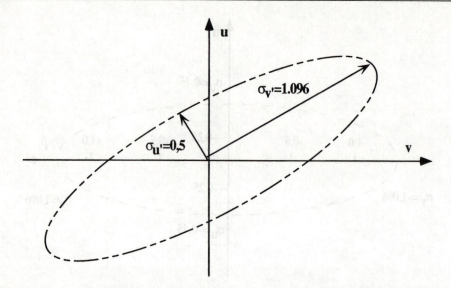

Figure 3.6: Example of rotation of the ellipse of the previous figure. The axes are u and v. $\sigma_{u'} = 0.500$ and $\sigma_{v'} = 1.o96$ are also indicated, as axes after the rotation. The ellypse is deformed because of covariance, which is different from zero.

One therefore writes in general the uncertainty matrix, indicating $\sigma^2_{u'v'}$, as $cov(u', v')$:

$$E = \begin{pmatrix} \sigma^2_{u'} & cov(u', v') \\ cov(u', v') & \sigma^2_{v'} \end{pmatrix} \tag{3.65}$$

For simplicity of notation, we often introduce *the correlation coefficient as well* (see figure 3.7):

$$\varrho = \frac{cov(u, v)}{\sigma_u \cdot \sigma_v} \tag{3.66}$$

and equation 3.47, valid for non correlation, can be seen in an extended

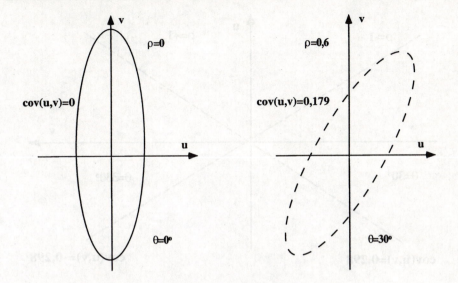

Figure 3.7: Values of parameter ϱ and $cov(u, v)$.

form which holds only if there is correlation:

$$\wp(u, v; \sigma_u, \sigma_v) = \frac{1}{2\pi\sigma_u\sigma_v} \frac{1}{\sqrt{1 - \varrho^2}} e^{-\frac{1}{2}\frac{1}{1-\varrho^2}\left(\frac{u^2}{\sigma_u^2} + \frac{v^2}{\sigma_v^2} - \frac{2\varrho uv}{\sigma_u\sigma_v}\right)} \quad (3.67)$$

The interpretation of coefficient ϱ is immediate; for example

$$\varrho = 0 \qquad cov(u, v) = 0 \qquad\qquad\qquad (3.68)$$
$$\varrho = +1 \qquad cov(u, v) \equiv \sigma_u \cdot \sigma_v \qquad\qquad (3.69)$$
$$\varrho = -1 \qquad cov(u, v) \equiv -\sigma_u \cdot \sigma_v \qquad\quad (3.70)$$

when $\varrho = 1$, the ellipse degenerates into a line as in figure 3.8. In the previous example $\varrho = -0.4$.

Sometimes, instead of the uncertainty matrix in equation 3.65, we

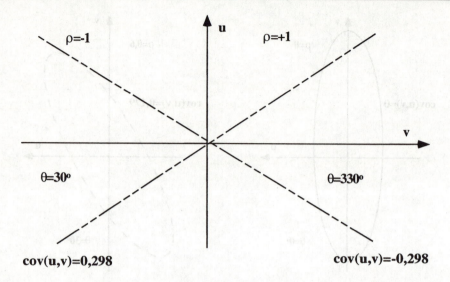

Figure 3.8: Values of parameter ϱ and $cov(u, v)$, when the ellipse degenerates into a line.

write, with definition 3.66, the reverse matrix:

$$M = E^{-1} = \frac{1}{(1 - \varrho^2)} \begin{pmatrix} \frac{1}{\sigma_u^2} & -\frac{\varrho}{\sigma_u \sigma_v} \\ -\frac{\varrho}{\sigma_u \sigma_v} & \frac{1}{\sigma_v^2} \end{pmatrix} \tag{3.71}$$

At this point we can come back to the general expression of the combination of errors 3.17 and see that it can be modified with the terms introduced here, into:

$$\sigma_x^2 = \begin{pmatrix} \frac{\partial x}{\partial u} & \frac{\partial x}{\partial v} \end{pmatrix} \begin{pmatrix} \sigma_u^2 & cov(u, v) \\ cov(u, v) & \sigma_v^2 \end{pmatrix} \begin{pmatrix} \frac{\partial x}{\partial u} \\ \frac{\partial x}{\partial v} \end{pmatrix} \tag{3.72}$$

briefly stated:

$$\sigma_x^2 = D^{-1} E D \tag{3.73}$$

We have therefore seen a succinct way of presenting the combination of uncertainties, which includes covariance.

3.4   Interpolation and combination of uncertainties

To determine an approximate value for uncertainty, we examined previously an analytical method. Here we present a method to determine an approximate value of a certain measurement which has not been directly measured. The uncertainty associated with it can be obtained using the method described in the previous chapter (3.1).and namely with the combination of uncertainties. Here we give a useful exercise on how to apply the method we have just studied.

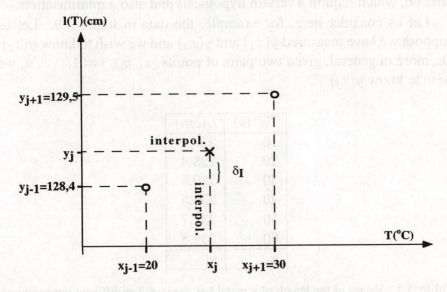

Figure 3.9: The plot shows the interpolation of a third measurement between two others.

In what follows we describe other methods more often used, for the estimate of accuracy parameters. These parameters, which are characteristic of experimental distributions, are used when they are compared with the parent ones to determine the best approximation to the data. In these cases we will use the minimisation procedure. In the methods presented in the two following chapters, we use, on the contrary, the recursive approximation applied directly to the ensemble of data, to obtain an approximation of a certain value from another one.

The two methods indicated here are interpolation and extrapolation. Both have an immediate graphic representation.

Here we describe the method of *interpolation*.

The method is useful both for data in plots and data in tables. It is of great use in the lack of a parent distribution or an experimental distribution. It is also easily applicable, with respect to other methods described later on, which require a certain hypothesis and also a minimisation.

Let us consider here, for example, the data in figure 3.9. Let us suppose we have measured $y(x_1)$ and $y(x_3)$ and we wish to know $y(x_2)$. Or, more in general, given two pairs of points (x_i, y_i), $i = 1, \dots, n$, we wish to know $y(x_j)$.

$T_i(°C)$	$l_i(cm)$
10	227.2
20	228.4
30	229.5
40	230.5
50	231.7
60	232.8
x_i	y_i

Table 3.1: Values of the length of a metal bar, measured at different temperatures. The values of elongation are exaggerated for the purposes of the example.

As an example, we can consider the measurement of the elongation of a metal bar as a function of its temperature [27]. The data are reported in table 3.1.

---

[27] Values are only indicative, and are exaggerated for the sake of the example.

If we wish, for example, to determine why $y_j \equiv l_j$ when $x_j \equiv m_j = 28°C$ we can use the following method, which is applicable only when (for each i)

$$\delta x = x_i - x_{i-1} = x_{i+1} - x_i \tag{3.74}$$

which corresponds to the condition that all x_i are equidistant and that the possible function of the plot, at least for one variable, can be approximated by a polinomial of any degree. Namely that:

$$y(x) \cong a + bx + cx^2 + dx^3 + \ldots \tag{3.75}$$

x_i	y_i	δ_i	δ_{II}	δ_{III}	δ_{IV}
x_1	y_1				
x_2	y_2	$\delta_i y_1$			
x_3	y_3	$\delta_i y_2$	$\delta_{II} y_1$		
x_4	y_4	$\delta_i y_3$	$\delta_{II} y_2$	$\delta_{III} y_1$	
x_5	y_5	$\delta_i y_4$	$\delta_{II} y_3$	$\delta_{III} y_2$	$\delta_{IV} y_1$

Table 3.2: Consecutive differences between interpolation values.

Reassessing the data in table 3.2 and calculating the differences defined as:

$$\delta_i y_1 = y_2 - y_1$$
$$\delta_i y_2 = y_3 - y_2$$
$$\delta_i y_3 = y_4 - y_3$$
$$\delta_i y_4 = y_5 - y_4$$

and

$$\delta_{II} y_1 = \delta_i y_2 - \delta_i y_1$$
$$\delta_{II} y_2 = \delta_i y_3 - \delta_i y_2$$
$$\delta_{II} y_3 = \delta_i y_4 - \delta_i y_3$$
$$etc. \ldots$$

we can take simple steps to obtain a general formula by induction. Since

$$y_2 = y_1 + \delta_i y_1 \equiv (y_1 + y_2 + y_3) \tag{3.76}$$

while

$$y_3 = \tag{3.77}$$
$$= y_2 + \delta_i y_2 =$$
$$= y_1 + \delta_i y_1 + \delta_i y_2 =$$
$$= y_1 + \delta_i y_1 + y_3 - y_2 =$$
$$= y_1 + \delta_i y_1 + y_3 - y_1 - \delta_i y_1 =$$
$$= y_1 + y_2 - y_1 + y_3 - y_1 - y_2 + y_1 =$$
$$= y_1 + y_2 - y_1 + y_3 - y_1 - y_2 + y_1 + (y_2 - y_2) =$$
$$= y_1 + \delta_i y_1 + (y_2 - y_1) + (y_3 - y_2) + (y_1 - y_2) =$$
$$= y_1 + \delta_i y_1 + \delta_i y_1 + \delta_i y_2 - (y_2 - y_1) =$$
$$= y_1 + 2\delta_i y_1 + \delta_i y_2 - \delta_i y_1 =$$
$$= y_1 + 2\delta_i y_1 + \delta_{II} y_1$$

and furthermore

$$y_4 = \tag{3.78}$$
$$= y_3 + \delta_i y_3 = \ldots =$$
$$= y_1 + 3\delta_i y_1 + 3\delta_{II} y_1 + \delta_{III} y_1$$

Then in general we can write:

$$y_k = y_1 + n\delta_i y_1 + \tag{3.79}$$
$$+ \frac{n(n-1)}{2!} \delta_{II} y_1 +$$
$$+ \frac{n(n-1)(n-2)}{3!} \delta_{III} y_1 + \ldots$$

with

$$n = \frac{x_j - x_{j-1}}{\delta x} \qquad (3.80)$$

Coming back to the example of the bar, we obtain table 3.3:

x_i	y_i	δ_i	δ_{II}	δ_{III}	δ_{IV}
10	227.2				
20	228.4	1.2			
30	229.5	1.1	-0.1		
40	230.5	1.0	-0.1	0	
50	231.7	1.2	$+0.2$	$+0.3$	$+0.3$
60	232.8	1.1	-0.1	-0.3	-0.6

Table 3.3: Consecutive differences between elongation values.

For example, to obtain, l_j for $T_j = 28°C$, we calculate

$$n = \frac{x_j - x_{j-1}}{\delta x} = \frac{28 - x_{j-1}}{\delta x} = \frac{28 - 20}{10} = 0.8 \qquad (3.81)$$

therefore

$$l_{j(a28)} =$$
$$= l_{j-1} + n\delta_i l_{j-1} + \frac{n(n-1)}{2!}\delta_{II} l_{j-1} +$$
$$+ \frac{n(n-1)(n-2)}{3!}\delta_{III} l_{j-1} + \cdots \cong$$
$$\cong 228.4 + 0.8(1.1) + \frac{0.8(-0.2)}{2}(-0.1) +$$
$$+ \frac{0.8(-0.2)(-1.2)}{3 \cdot 2 \cdot 1}(+0.3) + \cdots \cong$$
$$\cong 228.4 + 0.88 + 0.008 + 0.0096 + \cdots \cong 229.3$$

ν	χ_r^2	$P(\chi_r^2 \geq \chi_{r_o}^2)$
$\vdots$	$\vdots$	$\vdots$
9	0.463	0.90
9	0.598	0.80
$\vdots$	$\vdots$	$\vdots$

Table 3.4: Two values from the tables for the χ_r^2 parameter and the relative probabilities. It is necessary to interpolate between the two.

As can be seen from the previous example, the value has been obtained by consecutive approximation with the summing of progressively smaller terms of higher order. We could already have stated graphically that the point lies on a line, but in this way we have stated it more rigorously and with this expression we can also determine the associated uncertainty.

The uncertainty is calculated with the sum of the squares described in paragraph 3.1.1.

Here we give an example of data interpolation.

Example: The case of interpolation in the tables of χ^2

Interpolation between numbers in a table is a simple and frequent case in practice. It is particularly useful in applications, for example, for the reading of the tables of the χ^2 parameter which we will use in the following chapters for the evaluation of accuracy tests. It is also useful in evaluating confidence intervals and the corresponding areas indicated in the tables.

Let us suppose we read a table of the type 3.4 and have determined in the experiment a value of $\chi_{r_o}^2 = 0.49757$. Then the corresponding values are those indicated in table 3.5.

If we wish to know the corresponding value of $P(\chi_r^2) \geq P(\chi_{r_o}^2)$, we

$x_i \equiv \chi_r^2$	$y_i \equiv P(\chi_r^2 \geq \chi_{r_o}^2)$	δ_i
0.463	0.90	
0.598	0.80	-0.10

Table 3.5: Consecutive differences for the values of the χ_r^2 table.

have to determine the value by interpolation:

$$y_{interpolated} \equiv P(\chi_r^2 \geq 0.497) \equiv y_j \tag{3.82}$$

with

$$x_j = 0.497$$
$$\delta x = 0.598 - 0.463 = 0.135$$
$$n = \frac{(0.497 - 0.463)}{0.135} \cong 0.252$$

thus the value of y_j is :

$$y_j = y_1 + n\delta_i y_1 \equiv 0.90 + 0.252(-0.10) \cong$$
$$\cong 0.88 \equiv 88\% \tag{3.83}$$

At this point, an annotation, which will be useful after studying the chapter on accuracy tests, is in order. The χ_r^2 parameter has a large number of digits, but this does not confirm the accuracy of the corresponding measurements for the hypothesis for which it has been determined

Let us now return to the uncertainty to associate with y_j. This can be calculated according to the expressions and combinations discussed above. By repeating the calculations here, we can easily find that: $\Delta y_j \cong 0.01$. Therefore:

$$y_j \pm \Delta y_j = (0.88 \pm 0.01) \equiv (88 \pm 1)\% \tag{3.84}$$

In any case, this is an excellent result which gives 1% for the evaluation of the validity of the hypothesis.

To make more accurate evaluations, which are not necessary here, we should have tables with more accurate values of $P(\chi_r^2 \geq \chi_{r_o}^2)$. Therefore it is not necessary to determine the $\chi_{r_o}^2$ parameter with a number of significant digits larger than those given in the tables. Conversely, we should try to arrive at an accuracy of measurement such as to obtain the value with a number of significant digits similar to the one in the tables.

3.5 Extrapolation and combination of uncertainties

Compared to the previous case, extrapolation has the disadvantage of being less precise, as it does not allow approximations at higher orders. Furthermore, it is not easy to define the order and degree to which the procedure is reliable.

Let us suppose we have a certain datum x_j and that we want to know $y(x_j)$, not measured. As in the case of interpolation, the value can be approximated from a simple polynomial of degree 1. Therefore $y(x_j)$ at degree 1 is :

$$y_1(x_j) = a_1 + b_1\gamma \tag{3.85}$$

with a_1, b_1, coefficients to be determined, for example, by recursion. γ is given by:

$$\gamma = \frac{x_j - x_{j-1}}{x_{j+1} - x_{j-1}} \tag{3.86}$$

To make the approximation to obtain the coefficients, we can state that:

$$a_1 \equiv y_1(x_{j-1}) \tag{3.87}$$
$$b_1 \equiv y_1(x_{j+1}) - y_1(x_{j-1})$$

Increasing the degree to 2:

$$y_2(x_{j+1}) = a_2 + b_2\gamma + c_2\gamma^2 \tag{3.88}$$

which, solving the polynomial, as in equation 3.5:

$$y_2(x_{j+1-1}) = a_2 \equiv a_1 \tag{3.89}$$

$$\vdots$$

$$b_2 = b_1 - c_2 \tag{3.90}$$

$$c_2 = \frac{1}{2}(y(x_{j+2}) - 2y(x_{j+1}) + y(x_{j-1})) \tag{3.91}$$

Therefore in general, we can write:

$$y_2(x_{j+1}) = a_1 + b_1\gamma + c_2\gamma(\gamma - 1) + \cdots \tag{3.92}$$

with n points and degree $(n-1)^{th}$, indicating the various coefficients of higher order with a_i, instead of b, c, etc. :

$$\vdots$$

$$y_{n-1}(x_j) =$$
$$= a_1 + a_2\gamma + a_3\gamma(\gamma - 1) + a_4\gamma(\gamma - 1)(\gamma - 2) + \cdots =$$
$$= a_1 + \sum_{j=2}^{n}(a_j \prod_{k=1}^{j-1}(\gamma - k - 1)) \tag{3.93}$$

This is a general expression of the approximation of a datum with a polynomial.

By presenting the equations in this way they appear to be general. They are valid both for extrapolation and interpolation, for which in the previous paragraph we saw a very simple case of application in which all x_i were equidistant one from the others.

The case we are about to discuss here is not easy and practical to apply. The coefficients are obtained from a system of equations, starting from 3.5, but it is always best to determine them, at least in an experimental manner, from other measurements or from tables or in any case independently.

Example: The case of the number of windings in the experiment on magnetic induction

Let us consider the experiment on the induction of alternating current between two different coils, one inserted inside a larger one to which a known alternate current is applied. The experiment is described in Appendix A.6.

The current induced in the secondary circuit, namely in the smaller coil, V_s, depends linearly on the current applied to the primary circuit, V_p and on the number of windings it is made of. This is seen in the plot

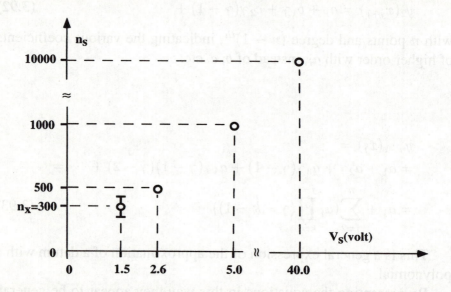

Figure 3.10: Number of windings of the secondary coil as a function of voltage V_s induced by the primary coil, in which the former is inserted. The value of the voltage corresponding to the number of unknown windings to be determined is indicated.

of the number of windings n, as a function of V_s for the same V_p, as in figure 3.10.

By inserting the coil with an unknown number of windings, n_x, this can be determined by measuring the corresponding induced voltage,

$V_s(n_x)$.

The plot in figure 3.10 indicates the corresponding point.

From the figure we see that the number n_x corresponds to the lowest possible voltage value. Furthermore, the dependence of V_s on $n_{windings}$, is linear, as expected. The curve passes through $V_s = 0$ to $n_{windings} = 0$, where there are no other experimental points between the unknown value we wish to determine and 0. Thus in this case the extrapolation process is necessary.

Actually, if we do not know parameters b and y_{j+1} (see the previous paragraph, identifying here y_j with n_x), it would not be possible to extrapolate without exploiting the fact that the relation between x, the voltages, and y, the number of turns, is linear, as well as the fact that the curve passes through the origin of the axes.

It is then possible to perform a sort of interpolation between $x_{j+1} = V_s = 0$ and $y_{j+1} = n_{windings} = 0$. It is to be noted that the index is $j+1$ and not $j-1$ because we are extrapolating beyond measured values x_j, y_j, even though these values numerically precede the others on the scale. $x_{j-...}, y_{j-...}$.

So we can write equation 3.85 in the form

$$\frac{y_{j+1} - y_j}{x_{j+1} - x_j} = \frac{y_j - y_{j-1}}{x_j - x_{j-1}} \tag{3.94}$$

Namely

$$\frac{n_{j+1} - n_x}{V_s(n_{j+1}) - V_s(n_x)} = \frac{n_x - 0}{V_s(n_x) - 0} \tag{3.95}$$

Therefore

$$n_x = \frac{V_s(n_x)n_{j+1}}{V_s(n_{j+1})} \tag{3.96}$$

From the data we have:

$$n_x = \frac{1.5 \cdot 500}{2.6} \cong 300 \, windings \tag{3.97}$$

Let us now combine the uncertainties for the determination of the uncertainty on n_x. We have to know the value of the uncertainty with which the number of windings is known. This is a typical case of assigning of a discrete uncertainty. The value in fact can only be 1 or multiples of 1. If there are no special manufacturing defects, and assuming that when the coils were wound, the number of windings was actually recorded, plausibly with a resolution equal to the first significant digit, it is reasonable to ascribe to it $\Delta n = 1$ for all the coils. On the other hand, if we have to determine an integer number, it is recommended to assume that the significant digits are not larger than those before the decimal point and that no other digits are to be considered.

On developing the calculations for the combination of uncertainties, we obtain

$$(n_x \pm \Delta n_x) \cong (300 \pm 50) windings \tag{3.98}$$

The result indicates that because of the extrapolation procedure the uncertainty is far larger than the one for the homogeneous quantities from which it was derived. It is therefore larger than the uncertainty given by the manufacturer on the number of windings turns. This confirms what we said at the beginning of this paragraph about the relative inaccuracy of the extrapolation method.

3.6   Mean standard deviation and the weighted mean

Owing to what we have studied so far on the combination of uncertainties, we shall now see what to do when we have more than one measurement and we wish to determine the uncertainty of a certain derived quantity: the arithmetic mean. The uncertainty to be associated with it is determined by the combination of uncertainties.

First of all, we have to give a definition of the arithmetic mean, by ascribing to each measurement *a weight* that is inversely proportional to the uncertainty with which it has been determined, a definition that corresponds to common sense. Correspondingly, we have a mean value of the parent distribution which is a combination of different distributions [29] each one with its own σ. Each value of x_i is multiplied by a coefficient w_i, which is inversely proportional to σ_i, i.e. for its inverse. In this way we indicate the new definition of the mean value:

$$\mu_w = \frac{\sum_{i=1}^{n} \frac{x_i}{\sigma_i^2}}{\sum_{i=1}^{n} \frac{1}{\sigma_i^2}} \tag{3.99}$$

which is more general than the one used so far. This is also called *weighted mean value* and reduces to the one introduced in the first chapter, when the weights to apply to all data are the same, i.e. the σ_i are all equal. Then $mu_w \equiv \mu$.

Likewise, we have a *weighted arithmetic mean* or simply *weighted mean*.

$$\overline{x} = \frac{\sum_i^n \frac{x_i}{s_i^2}}{\sum_i^n \frac{1}{s_i^2}} \tag{3.100}$$

Let us now determine uncertainty on the weighted mean. We distinguish two cases. Let us first discuss the one in which the uncertainty of each measurement contributing to the mean value, is the same for all, for example, since we are dealing with repeated measurements with the same instrument, in this case we are speaking about resolution. The other case occurs, for instance, when one takes the readings by measuring the same quantity with different instruments or on different scales of the same instrument.

Let us first examine uncertainty on the arithmetic mean, namely the weighted mean with all the $s_i = s_x$.

---

[29]The fact that the distribution has μ or σ equal to the mean of the other distributions, is often justified by the so-called emphcentral*limittheorem* which has not been introduced here because the fact that different homogeneous distributions can be combined together to obtain a like one is sufficiently intuitive.

Given a certain value of arithmetic mean of a series of independent measurements n all with $x_i \pm \Delta x_i$, $\overline{x}$, the uncertainty is given by the approximation of the corresponding value of the parent distribution and the experimental distribution:

$$\sigma_\mu = \sqrt{\sum_{i=1}^{n} (\frac{\partial \mu}{\partial x_i})^2 \sigma_i^2} \tag{3.101}$$

By developing the derivative:

$$\frac{\partial \mu}{\partial x_i} \cong \frac{\partial \overline{x}}{\partial x_i} =$$

$$= \frac{\partial}{\partial x_i} (\sum_{i=1}^{n} \frac{x_i}{n}) = \frac{\partial}{\partial x_i} (\frac{1}{n} \sum_{i=1}^{n} x_i) =$$

$$= \frac{1}{n} \frac{\partial}{\partial x_i} (\sum_{i=1}^{n} x_i) \tag{3.102}$$

The only term in this sum in which the derivative is not zero is the term $i - th$, whose derivative is 1. Therefore, for each x_i:

$$\frac{\partial \overline{x}}{\partial x_i} = \frac{1}{n} \cdot 1 \tag{3.103}$$

Coming back to equation 3.101:

$$\sigma_\mu = \sqrt{\sum_{i=1}^{n} (\frac{\partial \mu}{\partial x_i})^2 \sigma_i^2} \cong \sqrt{\sum_{i=1}^{n} (\frac{\partial \overline{x}}{\partial x_i})^2 s_i^2} =$$

$$= \sqrt{\sum_{i=1}^{n} (\frac{1}{n})^2 s_i^2} \tag{3.104}$$

Where the s_i replace the parent standard deviations if the $i - th$ measurement has been obtained with a series of measurements, or coincide with the resolutions or in any case with the uncertainties associated with the reading, if the x_i are single values directly measured. For simplicity's sake, let us consider the latter case. Thus the $s_i \equiv \Delta x_i$:

$$\sigma_\mu \cong \sqrt{\sum_{i=1}^{n}(\frac{1}{n})^2 \Delta x_i^2} \qquad (3.105)$$

In the former case we are considering all the $\sigma_i = \sigma_x$ and the $s_i = s_x \equiv \Delta x$. Then

$$\sigma_\mu \cong \sqrt{\sum_{i=1}^{n}(\frac{1}{n}\Delta x_i)^2} = \frac{\Delta x}{\sqrt{n}} \equiv \frac{s_x}{\sqrt{n}} = s_{\bar{x}} \qquad (3.106)$$

Let us now examine the case in which all the σ_i are different, that is, in practice, for example when the resolutions are different for different scales. Then,

$$\sigma_{\mu_w} = \sqrt{\sum_{i}^{n}(\frac{\partial \mu_w}{\partial x_i})^2 \sigma_i^2} \cong s_{\bar{x}_w} = \sqrt{\sum_{i=1}^{n}(\frac{\partial \bar{x}_w}{\partial x_i})^2 s_i^2} \qquad (3.107)$$

deriving the second member:

$$(\frac{\partial \bar{x}_w}{\partial x_i}) = \frac{\partial}{\partial x_i}(\frac{\sum_{i=1}^{n}\frac{x_i}{s_i^2}}{\sum_{i=1}^{n}\frac{1}{s_i^2}}) =$$

$$= \frac{1}{\sum_{i=1}^{n}\frac{1}{s_i^2}}\frac{\partial}{\partial x_i}\sum_{i=1}^{n}\frac{x_i}{s_i^2} =$$

$$= \frac{\frac{1}{s_i^2}}{\sum_{i=1}^{n}\frac{1}{s_i^2}} \qquad (3.108)$$

Therefore

$$\sigma_{\mu_w} \cong s_{\overline{x}_w} = \sqrt{\sum_{i=1}^{n} \left(\frac{\frac{1}{s_i^2}}{\sum_{i=1}^{n} \frac{1}{s_i^2}} \right)^2 s_i^2} =$$

$$= \frac{1}{\sqrt{\sum_{i=1}^{n} \left(\frac{1}{s_i^2} \right)}} = \frac{1}{\sqrt{\sum_{i=1}^{n} \left(\frac{1}{\Delta x_i^2} \right)}} \qquad (3.109)$$

which gives equation 3.106, with $\Delta x_i = constant \equiv \Delta x$.

Thus we have two expressions, from equations 3.106 and 3.109, for uncertainty on the mean:

$$\Delta \overline{x} = \frac{\Delta x}{\sqrt{n}} \qquad (3.110)$$

and

$$\Delta \overline{x}_w = \frac{1}{\sqrt{\sum_{i=1}^{n} \frac{1}{\Delta x_i^2}}} \qquad (3.111)$$

These two formulas are widely used in practice. But a word of caution is necessary in these definitions. It is meaningless to speak about the *mean error* or *mean uncertainty* in the light of these equations, even though their meanings are often confused. It is also to be underlined that the formula obtained above gives the uncertainty on the mean and it should not be misused in characterising the distributions. This is to say that it must not be confused with or replaced by the standard deviation of a distribution, which is often the case.

Example: The case of the uncertainty on the mean velocity in the experiment on coaxial cables

Let us now go back to the experiment on coaxial cables, in particular on the measurement of the lengths of the pieces of cable spliced together.

The idea is to determine the velocity of a signal along a coaxial cable for different time intervals t_i, corresponding to the relative delay between two signals, one of which passes through a cable to which pieces of length l_i are added to increase its length.

For this very experiment we can apply the two cases indicated above for the calculation of uncertainties on the means of the two quantities, $\Delta \bar{t}$ and $\Delta \bar{l}_w$.

As we have seen before, uncertainty Δl_i, increases with the increase in the value l_i, while Δt_i, which is constant, is the resolution as the reading error derived from the mark on the scales of the oscilloscope, equal to $5ns$ for each measurement.

We can therefore write:

$$\Delta \bar{t} = \frac{\Delta t}{\sqrt{n}} \qquad (3.112)$$

and

$$\Delta \bar{l}_w = \frac{1}{\sqrt{\sum_{i=1}^{n} \left(\frac{1}{\Delta l_i}\right)^2}} \qquad (3.113)$$

Supposing we can make $n = 20$ measurements for 20 different pieces of cable:

$$\Delta \bar{t} = \frac{5ns}{\sqrt{20}} \cong 1ns \qquad (3.114)$$

and, with the reading error of the meter $0.1cm$

$$\Delta \bar{l}_w = \frac{1}{\sqrt{\sum_{i=1}^{20} \frac{1}{\Delta l_i}}} \cong$$

$$\cong \frac{1}{\sqrt{\frac{1}{0.1} + \frac{1}{0.2} + \cdots + \frac{1}{0.5}}} \cong$$

$$\cong \frac{1}{\sqrt{10 + 5 + \cdots + 2}} \cong 0.2cm \qquad (3.115)$$

Therefore, by suitably weighting the measurements for their uncertainty, we are close enough to the accuracy indicated in the reading. If the weights are all equal this can be improved.

Even more different is the case for mean velocity, given by the ratio:

$$\bar{v} = \frac{\bar{l}_w}{\bar{t}} \tag{3.116}$$

with uncertainty

$$\Delta \bar{v} = \frac{1}{\bar{t}} \sqrt{\Delta \bar{l}_w^2 + \bar{v}^2 \Delta \bar{t}^2} \tag{3.117}$$

from the expression of the combination of uncertainty for the division.

An alternative is to consider the single values

$$v_i = \frac{l_i}{t_i} \tag{3.118}$$

with their uncertainty Δv_i, which is not constant. We can then calculate the weighted mean:

$$\bar{v}_w = \frac{\sum_{i=1}^{20} \left(\frac{v_i}{\Delta v_i^2} \right)}{\sum_{i=1}^{20} \left(\frac{1}{\Delta v_i^2} \right)} \tag{3.119}$$

and the associated uncertainty $\Delta \bar{v}_w$.

4   Accuracy tests

4.1   Hypotheses and the verification of accuracy

Up to now, we have been trying to identify practical instruments for the rigorous evaluation of the uncertainty associated with a given measurement. We will use what we have learned so far on the distributions of data and uncertainties to identify other, quite useful practical parameters to evaluate the quality of measurements.

At the end of this chapter we shall examine some applications in which the characteristic parameters of distributions are used to quantitatively evaluate the results or to determine quantities that characterise the sample of data. It is the case, for example, of the evaluation, in biomedical science, of the standard deviation of distributions, to derive quantities such as the dosage to apply to another sample and so on.

We can evaluate the characteristic parameters of the experimental distributions, but not those of the associated parent distributions without a rigorous method to identify them.

Therefore the initial problem of evaluation of the quality of measurements lies in the search for a suitable parent distribution.

Once we have done this, we will be able to use the characteristic parameters of the distributions to compare them. For example, we will be able to determine whether the standard deviations of two distributions are equal or similar within a certain interval of values and so on. This comparison may turn out to be sterile. If we wish to see if the standard deviation of an experimental distribution is or is not equal to the Gauss parent distribution, we will make use of the methods developed in this chapter. In the case of data that are counts of radioactive decay, the distribution may have a certain standard deviation equal to that of the Gauss distribution, but this does not necessarily represent the experimental distribution.

We need something more. From we have said thus far we can see the need to evaluate the adequacy of the parent distribution which most closely *resembles* the experimental distribution.

Students may be induced to say that they can roughly evaluate which parent distribution best *resembles* that of the experimental data. Referring back to chapter 2.2.4 we find that different parent distributions can have the same graphic form. Therefore not even this criterion [29] can be used; one similar to this is at the basis of the rigorous one that we describe here, so there is no need for a rough one.

4.2 Maximum likelihood

We can evaluate the *similarity* or *likelihood* between one or more distributions, for example, the parent and the experimental distributions, by evaluating the probability that the distribution is parent to an experimental distribution.

We therefore *have to formulate an a priori hypothesis* on the parent distribution. Therefore we can see that the evaluation of *likelihood* is the evaluation of an hypothesis.

By using probability for determining *likelihood*, we will avoid the danger of being misled by a small probability, since we are not able to quantify how small this will be. It is natural and intuitive to decide that the maximum probability corresponds to the *maximum likelihood*. Before comparing an experimental distribution and a corresponding parent one, we must discuss the distributions that will be used later on for the experimental ones.

Let us consider one or more distributions of the Gauss type for simplicity's sake. The probability function of an $i - th$ *set of data*, x_i to have a parent distribution with mean value μ and standard deviation σ_i,

---

[29] A *rough* criterion unfortunately still used in practice with sentences like: "... *a good experimentalist should be able to evaluate just with a ruler and a pencil...* ".

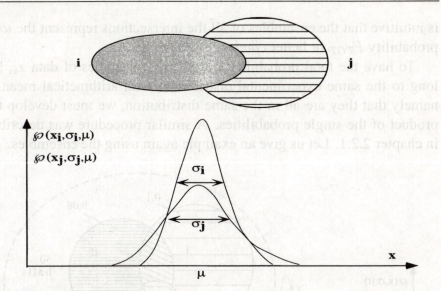

Figure 4.1: Illustration of intersection or overlapping of two ensembles. The overlapping of two different distributions is similarly represented below. $\wp(x_j, \sigma_j, \mu)$ and $\wp(x_i, \sigma_i, \mu)$.

is (chapter 2.2.3, equation 2.37):

$$\wp(x_i, \sigma_i, \mu) = \frac{1}{\sqrt{2\pi}\sigma_i} e^{-\frac{1}{2}\left(\frac{x_i - \mu}{\sigma_i}\right)^2} \tag{4.1}$$

For a different set of data, we have different probability functions, all with mean value μ [30].

By representing each distribution or set of data as an ensemble, it is plausible to hypothesize that there is an intersection between one or more ensembles, since each single distribution has the same mean value. This means that more than one ensemble will overlap (see figure 4.1). It

---

[30]This is plausible and can be demonstrated with the *theorem of the central limit* (see also chapter 3.6).

is intuitive that the ensembles of all the intersections represent the *total* probability P_{TOT} or better *combined* probability.

To have the total probability that all the ensembles of data x_i, belong to the same experimental distribution with arithmetical mean $\overline{x}$, namely that they are all in the same distribution, we must develop the product of the single probabilities. A similar procedure was described in chapter 2.2.1. Let us give an example again using the ensembles.

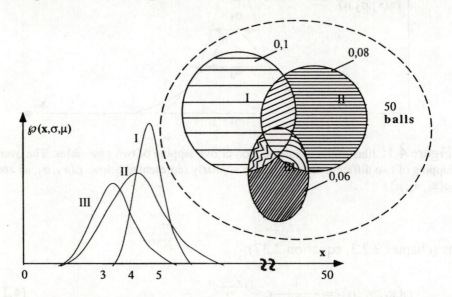

Figure 4.2: Illustration of the overlapping of three ensembles or of their *intersection* in the case of three red balls, extraction of the I, of the II and of the III. The overlapping of three different distributions is similarly represented below. $\wp(x_i, \sigma_i, \mu = 5)$ $\wp(x_{II}, \sigma_{II}, \mu = 4)$ $\wp(x_{III}, \sigma_{III}, \mu = 3)$.

Example: the ensembles of probability for coloured balls.

From a bag containing 50 coloured balls of which

5 red balls

10 black balls

15 white balls

20 of mixed colours

we wish to extract 3 red balls, the probability is

$$\wp(first, red) = \frac{5}{50}$$

$$\wp(second, red) = \frac{4}{49}$$

$$\wp(third, red) = \frac{3}{48}$$

which means

$$P_{TOT} =$$
$$= \wp(first, red)\wp(second, red)\wp(third, red) =$$
$$= \frac{5}{50}\frac{4}{49}\frac{3}{48} \cong 5.1 \cdot 10^{-4}$$

The ensemble resulting from the corresponding value of total probability is represented in figure 4.2.

Exercise: *red balls extracted repeatedly*

The student should repeat the previous example in the case in which after each extraction, the ball is put back into the bag.

It is therefore reasonable to write that the total probability function of the n distributions of data (each $i - th$ distributed with a probability function $\wp_i$ having the same mean value μ) is :

$$P(\mu) = \prod_{i=1}^{n}(\frac{1}{\sigma_i\sqrt{2\pi}}e^{-\frac{1}{2}(\frac{x_i-\mu}{\sigma_i})^2}) = (\frac{1}{\sigma_i\sqrt{2\pi}})^n e^{(\sum_{i=1}^{n}(-\frac{1}{2}(\frac{x_i-\mu}{\sigma_i})^2)}$$

$$(4.2)$$

On the basis of what we said earlier, we can now state that the maximum probability or *maximum likelihood*, is obtained when P_{TOT} is maximum. This happens, according to equation 4.2, if:

$$(\sum_{i=1}^{n}(-\frac{1}{2}(\frac{x_i - \mu}{\sigma_i})^2)) \tag{4.3}$$

is minimum. That is:

$$\frac{\partial}{\partial x}(\sum_{i=1}^{n}((\frac{x_i - \mu}{\sigma_i})^2)) = 0 \tag{4.4}$$

Therefore one has maximum likelihood between several distributions if 4.4 holds.

We can also make a similar case for several experimental distributions, by changing σ_i with s_i and μ with $\overline{x}$. Then 4.4 holds, provided that the approximation for $n \to \infty$ is verified.

4.3 The χ^2 parameter

We shall now use the concept of maximum likelihood to define an accuracy test using 4.4.

A simple and frequent application of the procedure of maximum likelihood is when we make more than one measurement and wish to know if the data follow *the theory*, or more properly, a *formulated hypothesis*.

In a very simple case we have a sample of data, which can be represented by a plot.

Example: measurements of length l of a metal bar as a function of temperature T.

Let us consider the measurements in table 4.1.

Let us suppose that we obtain from them a plot like the one in figure 4.3. We wish to see which curve best represents them.

$\bar{l}_i(cm)$	$T_i(°C)$
1.6	15
2.5	17
3.6	34
4.8	43
5.2	58
6.6	61
7.2	64
8.4	70
9.8	98

Table 4.1: Values of length l of a metal bar as a function of temperature T. Values are exaggerated to make the example clear.

We must remember that the single values represented in the plot are not to be confused with the data and the corresponding probability distributions. Each point in a graph can be obtained from a series of measurements which can *also*, but not necessarily, be distributed according to a Gauss distribution function.

In the plot we can, for example, indicate the arithmetical mean of the measurements $\bar{l}_i$. Each point represents an ensemble $i - th$ of data.

In the simple hypothesis that length l follows a linear function, in particular a simple polynomial of 1st degree, of temperature T, namely of the type

$$y = a + bx$$
$$l = a + bT$$

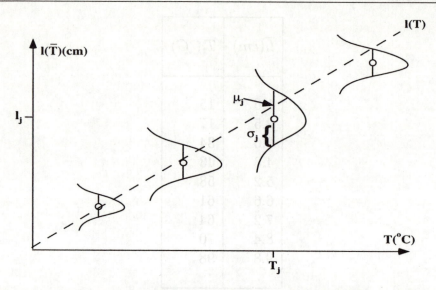

Figure 4.3: Graphic representation of the mean values of the lengths of a bar as a function of temperature. The different experimental distributions of different ensembles of data are represented and the function $l(T)$ of the hypothesis is indicated.

We can represent the function on the orthogonal plane with a line. Thus the graphic hypothesis is the equivalent of wondering if all mean values of lengths $\overline{l}_i$ fall on a line. To verifying the hypothesis in this case, is equivalent to ascribing to each experimental distribution of l_i, the $\wp_i(l)$, the corresponding parent distribution with mean value μ_{l_i} (to which $\overline{l}_i$ tends for $n \to \infty$) and standard deviation σ_{l_i} (to which s_i for $n \to \infty$) and looking for the maximum likelihood between each of the pairs of the distributions.

Obviously, the hypothesis of the simple polynomial binds each single parent distribution. Each one of them can be deduced from the other by an algebraic operation, namely by *shifting* along the line. This means that each value y_i can be obtained from the other one with either a sum or a subtraction of $a + bx_i$. As a consequence, we can consider the entire equation of the line as a unique hypothesis. Namely, we can consider all the points in the line as a unique parent distribution. With respect to this we have to verify the maximum likelihood.

The method of maximum likelihood is applied to each experimental distribution not with respect to the parent one, but one with respect to the others, with the condition that mean $\bar{l}$, obtained from the single means $\bar{l}_i$, is unique. The parent distribution, which is the one with the mean *theoretical* value, comes into the play because the true value of the Gauss function for maximum likelihood is the value of function $l(T)$, which represents the theoretical hypothesis, which is to say the equation of the line, in correspondence to abscissa T_i for which measurement l_{T_i} was made. In this case:

$$\wp_{T_i}(l_i) = \frac{1}{\sigma_{l_i}\sqrt{2\pi}}e^{-\frac{1}{2}(\frac{l_i - l_{T_i}}{\sigma_{l_i}})^2} \tag{4.5}$$

The condition of maximum likelihood of each experimental distribution of data (which is supposed to be Gauss-like), for each i, with respect to the others, is therefore (see 4.2):

$$P(\bar{l}) = \prod_{i=1}^{n} \wp_{T_i}(l_i) = \prod_{i=1}^{n}(\frac{1}{\sigma_{l_i}\sqrt{2\pi}}e^{-\frac{1}{2}\sum_{i=1}^{n}(\frac{l_i - l(T_i)}{\sigma_{l_i}})^2}) \tag{4.6}$$

In this equation, $l(T_i)$ is fixed, therefore the explicit independent variables in the equation of the line are the parameters, a and b. That is

$$P(\bar{l}) \equiv P(a,b) = \prod_{i=1}^{n}(\frac{1}{\sigma_{l_i}\sqrt{2\pi}}e^{-\frac{1}{2}\sum_{i=1}^{n}(\frac{l_i - l(T_i)}{\sigma_{l_i}})^2}) \tag{4.7}$$

Rewriting with

$$\Delta l_i = l_i - l(T_i) \tag{4.8}$$

then

$$P(\bar{l}) \equiv P(a,b) = \prod_{i=1}^{n}(\frac{1}{\sigma_{l_i}\sqrt{2\pi}}e^{-\frac{1}{2}\sum_{i=1}^{n}(\frac{\Delta l_i}{\sigma_{l_i}})^2}) \tag{4.9}$$

The condition of maximum likelihood, even in this case, is verified if $P(a, b)$ is maximum, that is if

$$\frac{1}{2} \sum_{i=1}^{n} (\frac{\Delta l_i}{\sigma_{l_i}})^2 \qquad (4.10)$$

is minimum, which happens when

$$\frac{\partial}{\partial l_i} \frac{1}{2} \sum_{i=1}^{n} (\frac{\Delta l_i}{\sigma_{l_i}})^2 = 0 \qquad (4.11)$$

We define

$$\chi^2 = \sum_{i=1}^{n} (\frac{\Delta l_i}{\sigma_{l_i}})^2 \qquad (4.12)$$

This parameter, which is widely used, will be discussed again further on, first as a parameter for minimisation, then as a parameter for the evaluation of the hypothesis, that is the maximum likelihood.

4.4   Minimisation

So far we have introduced maximum likelihood in order to minimise the function χ^2. This is also called *least square minimisation*, meaning the sum of equation 4.10.

To verify the condition of minimisation, it must hold that (in this simple case of the two variables, being $l(T) = a + bT_i$, $\chi^2(a, b)$):

$$\frac{\partial}{\partial a} \chi^2 = \frac{\partial}{\partial a} (\sum_{i=1}^{n} \frac{1}{\sigma_{l_i}^2} (\Delta l_i)^2) = 0 \qquad (4.13)$$

$$\frac{\partial}{\partial b} \chi^2 = \frac{\partial}{\partial b} (\sum_{i=1}^{n} \frac{1}{\sigma_{l_i}^2} (\Delta l_i)^2) = 0 \qquad (4.14)$$

Before we go ahead, let us suppose, for the sake of simplicity, that σ_{l_i} are constant. This may in fact be true, at least for a series or a sub-series of data. Therefore, with $\sigma_{l_i} \equiv \sigma_l$, equations 4.14 become:

$$\frac{\partial}{\partial a}\chi^2 = \frac{\partial}{\partial a}(\sum_{i=1}^{n} \frac{1}{\sigma_l^2}(\Delta l_i)^2) = 0 \tag{4.15}$$

$$\frac{\partial}{\partial b}\chi^2 = \frac{\partial}{\partial b}(\sum_{i=1}^{n} \frac{1}{\sigma_l^2}(\Delta l_i)^2) = 0 \tag{4.16}$$

Explicitely, we obtain:

$$\frac{\partial}{\partial a}\chi^2 = \frac{\partial}{\partial a}\frac{1}{\sigma_l^2}(\sum_{i=1}^{n}(l_{T_i} - a - bT_i)^2) =$$

$$= \frac{-2}{\sigma_l^2}(\sum_{i=1}^{n}(l_{T_i} - a - bT_i)) = 0 \tag{4.17}$$

$$\frac{\partial}{\partial b}\chi^2 = \frac{\partial}{\partial b}\frac{1}{\sigma_l^2}(\sum_{i=1}^{n}(l_{T_i} - a - bT_i)^2) =$$

$$= \frac{-2}{\sigma_l^2}(\sum_{i=1}^{n}T_i(l_{T_i} - a - bT_i)) = 0 \tag{4.18}$$

That is :

$$\sum_{i=1}^{n} l_{T_i} - \sum_{i=1}^{n} a - \sum_{i=1}^{n} bT_i = 0 \tag{4.19}$$

$$\sum_{i=1}^{n} T_i l_{T_i} - \sum_{i=1}^{n} aT_i - \sum_{i=1}^{n} b(T_i)^2 = 0 \tag{4.20}$$

or also:

$$\sum_{i=1}^{n} l_i = \sum_{i=1}^{n} a + \sum_{i=1}^{n} bT_i = an + \sum_{i=1}^{n} bT_i \qquad (4.21)$$

$$\sum_{i=1}^{n} T_i l_{T_i} = \sum_{i=1}^{n} aT_i + \sum_{i=1}^{n} b(T_i)^2 = a\sum_{i=1}^{n} T_i + b\sum_{i=1}^{n} (T_i)^2 \quad (4.22)$$

In the solution of the system of equations 4.21, it is easier to remember the general case for a system of this type:

$$y_1 = a_1 x_{11} + a_2 x_{12} + a_3 x_{13} + \ldots$$
$$y_2 = a_1 x_{21} + a_2 x_{22} + a_3 x_{23} + \ldots$$
$$y_3 = a_1 x_{31} + a_2 x_{32} + a_3 x_{33} + \ldots$$
$$\vdots \qquad (4.23)$$

In matrix form, for two variables:

$$\begin{pmatrix} y_1 \\ y_2 \end{pmatrix} = \begin{pmatrix} x_{11} & x_{12} \\ x_{21} & x_{22} \end{pmatrix} \begin{pmatrix} a_1 \\ a_2 \end{pmatrix} \qquad (4.24)$$

which in symbols is written:

$$Y = XA \qquad (4.25)$$

That is, the system of equations 4.21 is of the type:

$$a = \frac{1}{\delta}\left(\sum_{i=1}^{n} x_i^2 \sum_{i=1}^{n} y_i - \sum_{i=1}^{n} x_i y_i \sum_{i=1}^{n} x_i\right) \qquad (4.26)$$

$$b = \frac{1}{\delta}\left(n\sum_{i=1}^{n} x_i y_i - \sum_{i=1}^{n} x_i \sum_{i=1}^{n} y_i\right) \qquad (4.27)$$

The coefficients of equation 4.24 can be found with the method of Kramer:

$$a_1 = \frac{\begin{vmatrix} y_1 & x_{12} \\ y_2 & x_{22} \end{vmatrix}}{\begin{vmatrix} x_{11} & x_{12} \\ x_{21} & x_{22} \end{vmatrix}} \qquad (4.28)$$

where the determinants are indicated. For example, in the case in discussion, this holds for the system of equations 4.21:

$$\delta = \begin{vmatrix} x_{11}x_{12} \\ x_{21}x_{22} \end{vmatrix} = x_{11}x_{22} - x_{21}x_{12} \tag{4.29}$$

Explicitly

$$\delta = \sum_{i=1}^{n} x_i^2 n - \sum_{i=1}^{n} x_i \sum_{i=1}^{n} x_i = n \sum_{i=1}^{n} (x_i)^2 - \left(\sum_{i=1}^{n} x_i\right)^2 \tag{4.30}$$

and the corresponding coefficients and the variables of equations 4.21, are:

$$a_1 = ay_1 = \sum_{i=1}^{n} x_i y_i \tag{4.31}$$

$$a_2 = by_2 = \sum_{i=1}^{n} y_i \tag{4.32}$$

Therefore, by solving the system 4.21, with the method of Kramer:

$$a = \frac{\begin{vmatrix} \sum_{i=1}^{n} y_i & \sum_{i=1}^{n} x_i \\ \sum_{i=1}^{n} y_i x_i & \sum_{i=1}^{n} x_i^2 \end{vmatrix}}{\delta} \tag{4.33}$$

whilst

$$b = \frac{\begin{vmatrix} n & \sum_{i=1}^{n} y_i \\ \sum_{i=1}^{n} x_i & \sum_{i=1}^{n} x_i y_i \end{vmatrix}}{\delta} \tag{4.34}$$

The equations of system 4.27 are valid for the simple case of a first degree polynomial function with a known term and identical standard deviations of the distributions of each measurement.

We can now continue to present the more general case in which the standard deviations are not equal. In this case the system 4.14, becomes

$$\frac{\partial}{\partial a}\chi^2 = \frac{\partial}{\partial a}\sum_{i=1}^{n}(\frac{1}{\sigma_i^2}(y_i - a - bx_i)^2) = 0 \tag{4.35}$$

$$\frac{\partial}{\partial b}\chi^2 = \frac{\partial}{\partial b}\sum_{i=1}^{n}(\frac{1}{\sigma_i^2}(y_i - a - bx_i)^2) = 0 \tag{4.36}$$

whose solution gives the coefficients of the type (just for parameter b)

$$b = \frac{1}{\delta}\begin{vmatrix} \sum_{i=1}n\frac{y_i}{\sigma_i^2} & \sum_{i=1}n\frac{x_i}{\sigma_i^2} \\ \sum_{i=1}n\frac{x_iy_i}{\sigma_i^2} & \sum_{i=1}n\frac{x_i^2}{\sigma_i^2} \end{vmatrix} \tag{4.37}$$

This is the solution for coefficient b. It must be stressed that in the determinant the different standard deviations appear explicitly.

The complete solutions of the equation system 4.36, are:

$$a = \frac{1}{\delta}(\sum_{i=1}^{n}\frac{x_i^2}{\sigma_i^2}\sum_{i=1}^{n}\frac{y_i}{\sigma_i^2} - \sum_{i=1}^{n}\frac{x_i}{\sigma_i^2}\sum_{i=1}^{n}\frac{x_iy_i}{\sigma_i^2}) \tag{4.38}$$

$$b = \frac{1}{\delta}(\sum_{i=1}^{n}\frac{1}{\sigma_i^2}\sum_{i=1}^{n}\frac{x_iy_i}{\sigma_{the}^2} - \sum_{i=1}^{n}\frac{x_i}{\sigma_i^2}\sum_{i=1}^{n}\frac{y_i}{\sigma_i^2}) \tag{4.39}$$

$$\delta = \sum_{i=1}^{n}\frac{1}{\sigma_i^2}\sum_{i=1}^{n}\frac{x_i^2}{\sigma_i^2} - (\sum_{i=1}^{n}\frac{x_i}{\sigma_i})^2 \tag{4.40}$$

These expressions allow calculation of the value of the coefficients and of the determinant for the solution of the system of equations. It is important to remember that each coefficient must be evaluated with its own uncertainty. This is calculated with the expressions for the combination of uncertainties:

$$\sigma_a = \sqrt{\sum_{i=1}^{n} \sigma_i^2 \left(\frac{\partial a}{\partial y_i^2}\right)} \qquad (4.41)$$

$$\sigma_b = \sqrt{\sum_{i=1}^{n} \sigma_i^2 \left(\frac{\partial b}{\partial y_i^2}\right)} \qquad (4.42)$$

in which the derivatives are given by

$$\frac{\partial a}{\partial y_j} = \frac{1}{\delta}\left(\frac{1}{\sigma_j^2}\sum_{i=1}^{n}\frac{x_i^2}{\sigma_i^2} - \frac{x_j}{\sigma_j^2}\sum_{i=1}^{n}\frac{x_i}{x_j^2}\right) \qquad (4.43)$$

$$\frac{\partial b}{\partial y_j} = \frac{1}{\delta}\left(\frac{x_j}{\sigma_j^2}\sum_{i=1}^{n}\frac{1}{\sigma_i^2} - \frac{1}{\sigma_j^2}\sum_{i=1}^{n}\frac{x_i}{\sigma_j^2}\right) \qquad (4.44)$$

We used the index j only to indicate the single variable $j - th$ with respect to which we are making the derivation. Coming back to the special case of σ_i constant, equations 4.44, simplify to:

$$\frac{\partial a}{\partial y_j} = \frac{1}{\delta}\left(\sum_{i=1}^{n} x_i^2 - x_j\sum_{i=1}^{n} x_i\right) \qquad (4.45)$$

$$\frac{\partial b}{\partial y_j} = \frac{1}{\delta}\left(nx_j - \sum_{i=1}^{n} x_i\right) \qquad (4.46)$$

thus

$$\sigma_a = \sqrt{\sum_{j=1}^{n} \sigma^2 \left(\frac{\partial a}{\partial y_j}\right)^2} = \cdots \sqrt{\frac{\sigma^2}{\delta}\sum_{i=1}^{n} x_i^2} \qquad (4.47)$$

$$\sigma_b = \sqrt{\sum_{j=1}^{n} \sigma^2 \left(\frac{\partial b}{\partial y_j}\right)^2} = \cdots \sqrt{\frac{n\sigma^2}{\delta}} \qquad (4.48)$$

which, in general, if the $\sigma_i \neq constant$, are written as:

$$\sigma_a = \dots \sqrt{\frac{1}{\delta} \sum_{i=1}^{n} \frac{x_i^2}{\sigma_i^2}} \qquad (4.49)$$

$$\sigma_b = \dots \sqrt{\frac{1}{\delta} \sum_{i=1}^{n} \frac{1}{\sigma_i^2}} \qquad (4.50)$$

To be borne in mind is that σ_i are the standard deviations of the parent probability distributions Gauss function for each single experimental distribution of data $i - th$ or, in our example, σ_{l_i}. To determine the uncertainty on the coefficients we can use equations 4.50, simply by identifying the single σ_i with the Δ_i, meant as instrumental uncertainties. Otherwise, we can determine, for each single measurement, the standard deviation s_i of the $i - th$ experimental distributions of each given $i - th$, if these data have been obtained, for example, as an arithmetical mean of its distribution of m repeated measurements. This holds with the justification that $\sigma \cong s$. In this case, since each σ_i is given by:

$$\sigma_i \equiv \sigma_{y_i} = \sqrt{\lim_{m \to \infty} \frac{1}{m} \sum_{k=1}^{m} (y_k - \mu)^2} \qquad (4.51)$$

by identifying μ (see figure 4.3), true value, with the theoretical value $y(x_i) = a + bx_i$, which is, in turn, the same for each k, namely $y(x_k) = a + bx_k$:

$$\sigma_{y_i} = \sqrt{\lim_{m \to \infty} \frac{1}{m} \sum_{k=1}^{m} (y_k - a - bx_k)^2} \qquad (4.52)$$

which in the approximation $s \cong \sigma$, translates into:

$$\sigma_{y_i} \cong s_{y_i} = \sqrt{\frac{1}{m-2} \sum_{k=1}^{m} (y_k - a - bx_k)^2} \tag{4.53}$$

This is the expression for the experimental standard deviation of each point. The coefficient $\frac{1}{n-1}$, of the general definition, is replaced by $\frac{1}{m-2}$. With respect to the first case, in fact, here the arithmetical mean is unknown. Here the demonstration we presented in introducing the numerical factor -1 is to be recalled. This was due to the fact that in that case at least one datum, the arithmetical mean, was known as a function of the others, by the equation defining it. Here, instead of the arithmetical mean, we know function $y(x_j) = a + bx_j$ which depends upon two coefficients, which can be determined with the two independent equations 4.27. The factor at the denominator is, in general, $m - \alpha$ where $\alpha = q + 1$ with q order of the polynomial of the function $y(x)$. In this simple case, in which the function is a linear equation with 2 coefficients a and b, $q = 1$ holds and the known term is 1, therefore $\alpha = 1 + 1 = 2$.

The expression 4.53 clarifies the interdependence of the value of the variables on a unique theoretical hypothesis, as well as on their uncertainty. Indeed, in the expression of the estimate of uncertainty, s_i for y_i the number of measurements and the degree of the polynomial appears.

In the simple case in which the σ_i are constant, we have

$$\sigma_y \cong s = \sqrt{\frac{1}{m-2} \sum_{j=1}^{m} (y_j - a - bx_j)^2} \tag{4.54}$$

Equations 4.50 for the errors in the two coefficients become:

$$\sigma_a \cong \sqrt{\frac{s^2}{\delta} \sum_{i=1}^{n} x_i^2} \cong \frac{s}{\sqrt{\delta}} \sqrt{\sum_{i=1}^{n} x_i^2} \tag{4.55}$$

$$\sigma_b \cong \sqrt{\frac{ns^2}{\delta}} = \frac{s}{\sqrt{\delta}} \sqrt{n} \tag{4.56}$$

We have therefore obtained the values of the coefficients and the relative errors with the minimisation method for the linear expression which best approximates the data.

4.5   The χ^2 accuracy test

So far we have discussed a method for the determination of the parameters and errors for a linear function through the minimisation of the χ^2. By using the parameter χ^2 we can evaluate the validity of the hypothesis expressed analytically by the linear function.

Let us go back to equation 4.54, bearing in mind the definition of χ^2, in equation 4.12

$$\chi^2 = \sum_{i=1}^{n}(\frac{\Delta y_i}{\sigma_i})^2 \qquad (4.57)$$

which, with $\sigma_i = constant \equiv \sigma$, becomes

$$\chi^2 = \sum_{i=1}^{n}(\frac{\Delta y_i}{\sigma})^2 \qquad (4.58)$$

But from 4.54, we have:

$$\sigma_i \equiv \sigma \cong s = \sqrt{\frac{1}{m-2}\sum_{j=1}^{m}(\Delta y_j)^2} \qquad (4.59)$$

Since σ is constant, we can also write that [31]:

$$\sigma_i \equiv \sigma \cong s = \sqrt{\frac{1}{m-2}\sum_{i=1}^{n}(\Delta y_i)^2} \qquad (4.60)$$

Coming back to 4.59:

$$\chi^2 \cong \frac{1}{\frac{1}{m-2}(\sum_{j=1}^{m}(\Delta y_j)^2)}\left(\sum_{j=1}^{m}(\Delta y_j)^2\right) = m - 2 \equiv \nu \qquad (4.61)$$

In this expression we see explicitly the number of degrees of freedom ν. For the time being, we emphasize that parameter χ^2 must not be confused with the number of the degrees of freedom. We will discuss equation 4.61 again. Before using χ^2, we shall discuss in details the parameter ν.

4.5.1   The number of degrees of freedom

The parameter ν is *the number of degrees of freedom* [32], the number of independent trials. In this specific case it is the number of measurements that contributes independently to the evaluation of the agreement of the theoretical hypothesis with experimental data. It is the number of trials with which the maximum likelihood is verified. These may be obtained by n measurements, but when explicating the theoretical hypothesis with α known parameters, then only $n - \alpha$ determines the number of really independent measurements.

---

[31]This argument may appear to be simply a new definition of the indexes. But in reality it is crucial in what follows. It is also the basis of the theorem of the central limit. Here we are assuming that, s being constant, the standard deviations $s_i \equiv s$, for each distribution, tend towards a unique s when $n \to m$. This happens when $n \to m \to \infty$.

[32]This terminology, which comes from algebra and is widely applied in analytical mechanics, may be misleading. The student may, in fact, carry over the prejudice of definitions given for specific examples.

Example: linear equation and degrees of freedom.

In the simple case of a linear equation

$$y(x_i) = a + bx_i \tag{4.62}$$

the measurements are n, the independent parameters a and b, then $\alpha = 2$, $\nu = n - 2$. Even if the known term does not appear explicitly in equation:

$$y(x_i) = bx_i \tag{4.63}$$

the interception of the line with the origin of the coordinates, even if it is equal to 0, must be considered in the same manner as the known term a which therefore does not have to be determined with the procedure described earlier, since it is defined by the experimental measurements. In this case one of the α parameters is determined not by equations 4.40, but independently with the measurements. In the value of ν, degrees of freedom equal to $\nu = n - \alpha$, since the point at 0 appears in the quantity n, it must not appear again in the quantity α otherwise it will be counted twice, thus underestimating the value of the degrees of freedom. There may be cases in which one of the parameters is determined from other theoretical hypothesis. In any case, they always enter into the calculation of total ν, as n and not as α.

Therefore n is not always the sole quantity of measurements. Similarly, we must not confuse the cases in which we leave the interval of the parameters quite wide to verify the theoretical hypothesis. When we calculate χ^2 we must take into account that the degrees of freedom are always $\nu = n - \alpha$ and in α the number of parameters which were made vary within large intervals is also included.

4.5.2 The parameter χ_r^2

Although we often speak about χ^2 as a characteristic parameter of the quality of likelihood, it is best to underline the fact that it is not the value of the parameter itself which allows an immediate evaluation of the adequacy of the theoretical hypothesis. In the case of the linear function, in which the known parameters are the coefficients of the polynomial equation, the degree of the polynomial, as well as the number of measurements, may change the value of χ^2, as can be seen in equation 4.61. It is necessary to specify the value of parameter χ^2 and the number of degrees of freedom ν. The two data come together when evaluating parameter χ^2 *reduced*, defined as:

$$\chi_r^2 = \frac{\chi^2}{\nu} \tag{4.64}$$

evaluation of the likelihood via the value of χ_r^2 is immediate. In fact if $\chi_r^2 \cong 1$, there is good agreement between theoretical hypothesis and experimental values. Furthermore, the farther χ_r^2 is from unity, the worse is the agreement between $y_j^{experimental}$ and $y_j^{theoretical}$. This is explicitly indicated in 4.61. We explain how to perform the quantitative evaluation in paragraph 4.5.3.

It is worth noting that equations 4.60 and 4.61 give an approximate expression, which is in any case valid in the hypothesis of $s \to \sigma$ for $n \to \infty$ and σ_j are the Gauss functions. It is to be remembered that this is the hypothesis of maximum likelihood.

Exercise: χ^2 *for the linear expression*

For the values of l and of T of table 4.1, with the reading uncertainty on temperature equal to $0.1°C$:

-make up a table with the corresponding x_i, y_i, Δy_i, x_i^2, y_i^2, of the specific case;

- *determine and write in the table, a, b, c, δ, σ_a, σ_b, s;*

- *determine ν, χ^2, χ_r^2.*

Minimisation for any order

Before going into greater detail and proceeding to show the practical applications of χ_r^2, we shall generalise here the procedure we followed in the previous chapters, to the case of minimisation for a polynomial of any degree q. This will bring us to a redefinition of χ_r^2.

The linear function can have a degree $l = 1, \ldots , q$. As for the linear equation for $l = 1$, we have:

$$y(x) = a + bx \qquad\qquad q = 1 \qquad term \qquad constant = 1$$
$$y(x) = a + bx + cx^2 \qquad q = 2 \qquad constant \qquad term = 1$$
$$y(x) = a + bx + \cdots + x^q \qquad q \qquad constant \qquad term = 1$$

Therefore, as before we have:

$$\nu = m - 2 = \qquad m - (1 + 1) \qquad q = 1 \qquad (4.65)$$

now

$$\nu = n - \alpha = n - q - 1 \qquad\qquad (4.66)$$

Coming back to equation 4.61, the experimental standard deviation is :

$$\sigma_{y_i} \cong s = \sqrt{\frac{1}{n-2} \sum_{i=1}^{n} (y_i - a - bx_i)^2} \qquad (4.67)$$

Which can be generalised in:

$$s = \sqrt{\frac{1}{n-q-1} \sum_{i=1}^{n} (\Delta y_i)^2} \qquad (4.68)$$

Keeping in mind the definition of χ^2, with $\sigma_i = constant = \sigma$, equation 4.59 and multiplying and dividing the second term of equation 4.68 by σ^2:

$$s^2 \cong \frac{\sigma^2}{\sigma^2} \frac{1}{n-q-1} \sum_{i=1}^{n} (\Delta y_i)^2 = \tag{4.69}$$

$$= \frac{\sigma^2}{n-q-1} \chi^2 \tag{4.70}$$

Therefore

$$s \cong \sqrt{\frac{1}{\nu} \sigma^2 \chi^2} \tag{4.71}$$

This equation 4.71, is to be compared with the definition of the χ_r^2, given in 4.64 and actually becomes:

$$\frac{s^2}{\sigma^2} \cong \frac{\chi^2}{\nu} \equiv \chi_r^2 \tag{4.72}$$

The same can be interpreted as a definition of χ_r^2 and this is often found in textbooks. It is a relation obtained for a very common case, but not necessarily the most general one, which is the minimisation of linear expressions to any order of the linear function.

Coming back to the comment made in paragraph 4.5, the experimental standard deviation s is the standard deviation of the sample of data. In the hypothesis of Gauss distributions, the standard deviations, $\sigma_i \equiv \sigma$, all tend to one σ, the mean of the distributions. In this case relation 4.72 still holds if one has a number $n \to \infty$ of distributions of $m \to \infty$ of data, for the mean experimental standard deviation $s_{\bar{x}}$ and σ_μ mean of the parent one. Therefore we can write:

$$\frac{s_{\bar{x}}^2}{\sigma_\mu^2} \equiv \frac{\chi^2}{\nu} \equiv \chi_r^2 \tag{4.73}$$

Example: The case of ν for a Gauss function in the photoresistor experiment

We mentioned before the experiment on the photoresistor made out of a silicon semiconductor. In this case we measure the value of the resistance by varying the intensity of incident light. This is performed by rotating a polariser between the lamp and the photoresistor.

Figure 4.4 shows an experimental curve obtained by rotating the polarising filter from 0° to 180°. The curve resembles a symmetric bell. Here we wish to see if a Gauss function can approximate the shape [33].

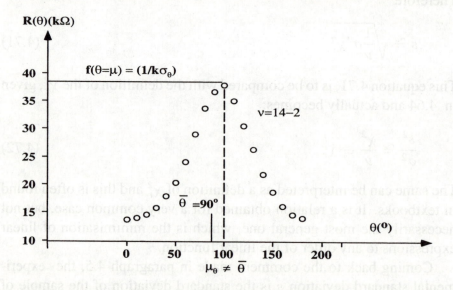

Figure 4.4: Values of resistance as a function of the degree of rotation of the polarising filter. The curve may be approximated with a function similar to a Gauss one. The parameters for the evaluation of the number of degrees of freedom are also indicated. The curve is not symmetric around the experimental arithmetical mean for an error of zero.

---

[33]The physical process at the basis of the behaviour of the photoresistor does not

The Gauss function has two parameters, μ and σ of the type:

$$f(x, \mu, \sigma) = \frac{1}{\sigma} e^{-\frac{(x-\mu)^2}{2\sigma^2}} \qquad (4.74)$$

We shall proceed by verifying the experimental data. We can operate in different ways and, depending on which one is adopted, we have to pay great attention to the fact that the number of degrees of freedom is different in the two cases.

Let us consider the function as it is and just add a coefficient k such that the scales between the values of the function calculated and the ones measured, both in angles and in resistance values, correspond. We can write:

$$f(x, \mu, \sigma) = \frac{1}{\sigma \cdot k} e^{-\frac{(x-\mu)^2}{2\sigma^2}} \qquad (4.75)$$

It is to be noted here that in this way the coefficient has a role similar to the one *of normalisation* introduced when we stated that the integral of the Gauss function represents total probability. Another important remark is that the coefficient can be written either to the numerator, or similarly, as in the example, to the denominator. Furthermore, the coefficient, like all the other parameters, has its own unit of measurement, which is needed to make $f(x)$ homogeneous.

Therefore in reality the function has 3 parameters. We shall see, however, that there are relations between them and that these relations allow us to correctly determine the number of degrees of freedom.

When $x = \mu$, $f(x)$ holds:

$$f(x = \mu) = \frac{1}{k\sigma} \qquad (4.76)$$

---

really legitimise this hypothesis. In fact, the curve should be exactly separable into two exponential functions of opposite sign in the exponent. The case of the Gauss function described here is given only as an example of the application to verify an hypothesis by testing.

Therefore, if we wish to determine the number of degrees of freedom, ν, from the number of n, then we will have to count

$$\nu = n - (3 - constraints) \tag{4.77}$$

Here by *constraints* we mean all the relations, that is, the equations, which bind or *constrain* the parameters one to the other. In this case, since there is only one relation, equation 4.76, between two parameters, then the constraints are 1 and one has:

$$\nu = n - (3 - 1) = n - 2 \tag{4.78}$$

We are advancing the hypothesis that parameters μ and σ are *free*. We looked for the maximum likelihood of the curve of the data with the function $f(x)$ without defining the parameters a priori. In reality, the experimental conditions do not allow any variation, neither for μ nor for σ. For example, it would not make sense to define $\mu = 10°$, because clearly one would never obtain a good value of χ^2 (see the discussion that follows, to understand what we mean by *good value of* χ^2), since this value is too far from the experimental data. However it is advisable that the χ_r^2 would in any case be obtained by dividing χ^2 by the greatest number of degrees of freedom possible, ν, because this would imply both having many measurements and few *free* parameters, thus little uncertainty on the function. At the same time this is not recommendable, because too few parameters make the search for the best agreement more difficult.

Sometimes experimental data may help, for example, by defining μ as known from the data and coincident with the mean value [34], then $\bar{\vartheta}$, the mean rotation angle of the polariser. In this case:

$$\nu = n - 3 - 2 = n - 1 \tag{4.79}$$

---

[34]It is to be noted here that in this example μ and $\bar{x}$ and σ and s have been confused throughout, meaning that the evaluation and the relations of the parameters of f(x) are actually the experimental ones, but for simplicity's sake we indicate only the theoretical values.

It is to be noticed that the choice of $\mu \cong \bar{\vartheta}$ is inadequate because the Gauss function, $f(\mu)$, would be maximum, while $f(\bar{\vartheta})$ would not correspond to the maximum value of R. Furthermore, the same equation 4.76, would no longer be valid. Therefore it is highly recommended to consider the experimental data, R_{max}, and therefore read the corresponding value of $\vartheta \cong \mu$.

To avoid complicating this example further let us suppose that R_{max} has been detected from the measurements, so equation 4.79 holds. In this case the function would keep only one free parameter, but it would be clear that it is difficult to find an agreement with the data by varying σ, which means the width of the curve, but not its maximum or its centre.

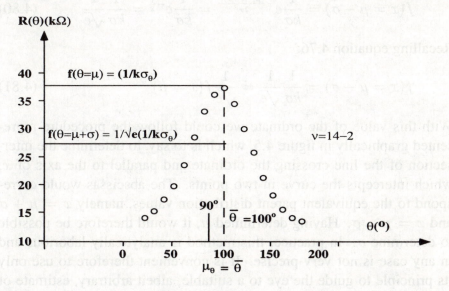

Figure 4.5: Values of resistance as a function of the degree of rotation of the polariser. The curve may appear approximated with a function similar to a Gauss one. The parameters for evaluating the number of degrees of freedom are indicated. The curve is symmetric around the experimental arithmetic mean.

To be noted once again that in any case the curve is not centred around the experimental values. This indicates an offset error 0 in scale ϑ, as discussed in the previous example, which makes agreement even more difficult.

Here is a case in which the degrees of freedom should not be diminished; the parameters must be free to vary, instead of using experimental information.

Another case is verified if, for example, we assume that the curve is well centred, as in figure 4.5: we could then consider $\bar{\vartheta}$ (meant as an analytical parameter and not a statistical one) as the best estimate of μ.

With this value we could have a further reliable constraint for σ. In fact:

$$f(x = \mu - \sigma) = \frac{1}{k\sigma}e^{-\frac{(\mu-\sigma-\mu)^2}{2\sigma^2}} = \frac{1}{k\sigma}e^{-\frac{1}{2}} = \frac{1}{k\sigma}\frac{1}{\sqrt{e}} \qquad (4.80)$$

Recalling equation 4.76:

$$f(x = \mu - \sigma) = \frac{1}{k\sigma}\frac{1}{\sqrt{e}} = \frac{1}{\sqrt{e}}f(x = \mu) \qquad (4.81)$$

With this value of the ordinate we could follow the procedure represented graphically in figure 4.5, which is to say, to determine the intersection of the line crossing the ordinate and parallel to the axis of x, which intercepts the curve in two points. The abscissas would correspond to the equivalent parent distribution values, namely $x = \mu + \sigma$ and $x = \mu - \sigma$. Having determined μ, it would therefore be possible to determine σ. In practice, this method is analytically laborious and in any case is not very precise. It is convenient therefore to use only its principle to guide the eye to a suitable, albeit arbitrary, estimate of the parameter. In this case, having determined σ from μ, it would be legitimate even to write:

$$\nu = n - 3 - 3 = n \qquad (4.82)$$

That is, the curve would be determined entirely experimentally, having fixed μ and k from the data and having deduced σ from them. It is to be underlined that this case is not recommendable, since it would void the minimisation procedure, reducing it to a mere analytical verification of the behaviour of the experimental function. The procedure illustrated is a limit case used here only to explain the use of parameter ν.

4.5.3 The χ^2 probability function

In order to use the parameter χ_r^2 rigorously, it is necessary to introduce a probability distribution function, similar to the Gauss, Poisson and Binomial distribution functions. In this way we can translate considerations on the evaluation of the quality of the hypothesis and the data related to it, in evaluations of the probability that the sample of data has that χ_r^2.

In order to do this we assume that *in the whole universe* there is an infinite number of trials and samples of data which can be collected within that specific experimental situation. We assume that each of these can be matched with, or *with respect to* a certain hypothesis which is analytically expressed, for example, as a linear function of a certain order.

We must remember that, in the hypothesis of maximum likelihood, the requirement is that in any case the samples must all be well distributed according to a Gauss probability function and they must be composed of all independent data.

We define a probability function that can be intuitively obtained from the combination of the probability functions of each measurement and the relative hypothesis to obtain that χ^2. In general, in the probability function of χ^2, the dependence on variable χ^2 and parameter ν is explicit:

$$\wp(\chi^2, \nu) = \frac{\sqrt{(\chi^2)^{(\nu-2)}}e^{-\frac{\chi^2}{2}}}{2^{\frac{\nu}{2}}\Gamma(\frac{\nu}{2})} \tag{4.83}$$

The graphic representation of this function makes clearer the relevance of parameter ν: if ν changes, it also makes a great change in the shape of the curve. For example, for $\nu = 1$, in figure 4.6, the curve has the shape of an hyperbole, with very high values of probability for very small values of χ^2. This means that for a small number of degrees of freedom the probability of obtaining a large χ^2 is very low.

The problem of determining the reliability of the result of a test is thoroughly discussed in the paragraph on confidence intervals. For now, we shall refer only to the use of the value of probability, not to its determination.

The student must use here the intuitive concept as already indicated in chapter 2, for the use of the definition of probability.

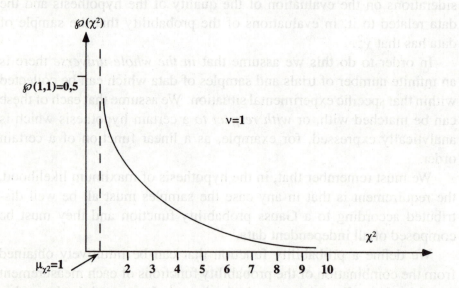

Figure 4.6: The curves of probability distribution functions, $\wp(\chi^2, \nu)$, for $\nu = 1$. The value of $\wp(\chi^2.1)$ for $\chi^2 = 1$ is indicated.

In the case discussed, low probability means that *in the whole universe* it is a minority. This means that *in the whole universe* there is a very low probability either that some other experimentalist obtains the same errors and/or the same results by repeating the trials in similar conditions, or that some other hypothesis happens to be correct with that data sample, or even that both these situations take place simultaneously. From this point of view, the value of probability is the quantification of the possibility of belonging or not to a population sample that is experimented (or *hypothesized*) as being large with respect to the *universe*, meant as the total number of population samples. It is therefore the concept of minority or *limited minority* or *majority* or *limited majority* or *absolute majority* or *relative majority*, however reached, that the student must keep in mind in evaluating the validity of the hypothesis in the probabilistic sense. For example, in the case of figure 4.6, the

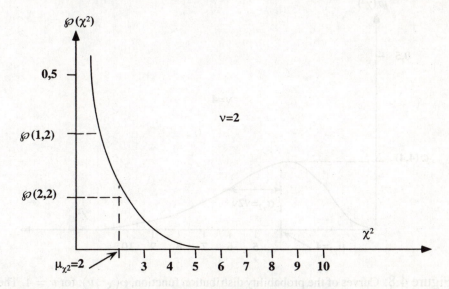

Figure 4.7: Curves of the probability distribution function, $\wp(\chi^2, \nu)$, for $\nu = 2$. The value of $\wp(\chi^2, 2)$ for $\chi^2 = 1$ and the value of $\wp(\chi^2, 2)$ for $\chi^2 = 2$ is indicated.

probability that χ^2 equals 1 is 50%, fairly probable.

We can see from the same example that the value of χ^2 cannot be taken by itself, without the determination of the number of degrees of freedom, for the evaluation of the validity of the hypothesis, nor *must*

χ^2 *be equal to* 1. In fact, for values of χ^2 lower than 1 the probability is even greater.

It must be stressed that not even the value of probability alone can be used for the evaluation, if we wish to compare different samples and different degrees of freedom. For example, figure 4.7, shows the probability value in the ordinates, which can assume values in a smaller interval, from 0.5, that is 50%, to 0. The value of the probability for the same value of $\chi^2 = 1$, as in the previous example, is also indicated. This value of $\wp(1,2)$ is smaller than the $\wp(1,1)$ of the previous case.

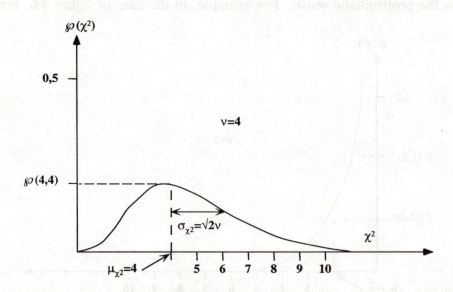

Figure 4.8: Curves of the probability distribution function, $\wp(\chi^2, \nu)$, for $\nu = 4$. The value of $\wp(\chi^2, 4)$ for $\chi^2 = 4$ (mean value) is indicated.

In figure 4.8, the curve indicates that the behaviour is no longer monotone as in the first two simple cases, but there is rather a maximum. This is not for increasingly smaller values of χ^2, as before, but for values close to the number of degrees of freedom though not exactly equal to them. That is, the probability of χ^2 is high for values close to the number of degrees of freedom. Figure 4.9 shows even more clearly that the closeness of the maximum around the value of χ^2, is not very significant. For $\nu = 10$, the curve has an ordinate with a maximum value still lower than 50% [35].

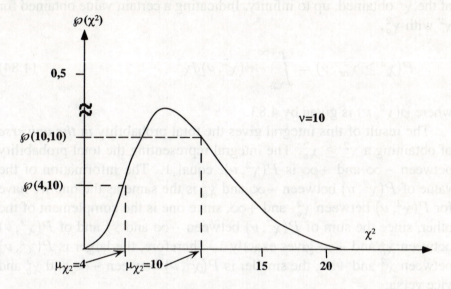

Figure 4.9: Curve of the probability distribution function, $\wp(\chi^2, \nu)$, for $\nu = 10$. The value of $\wp(\chi^2, 10)$ for $\chi^2 = 4$ and the value of $\wp(\chi^2, 10)$ for $\chi^2 = 10$ are indicated. They correspond to the mean values in $\nu = 4$ and $\nu = 10$

Use of $\wp(\chi^2)$

To have a method of evaluation that makes use of χ^2, which is of rapid use in practice, we can calculate the total probability $P(\chi^2, \nu)$,

---

[35]It is important to note that this test is not applicable for any value whatsoever of χ^2, nor for any value of ν. In fact, the $\wp(\chi^2, \nu)$ may be too small for the application of a suitable probabilistic criterion.

instead of calculating the sole probability $\wp(\chi^2, \nu)$, which is the integral of $\wp(\chi^2, \nu)$.

Conventionally, it is common to use a more significant integral, one that corresponds to the area below the curve of $\wp(\chi^2, \nu)$, from the value of the χ^2 obtained, up to infinity. Indicating a certain value obtained for χ^2 with χ_o^2,

$$P(\chi^2 \geq \chi_{or}^2, \nu) = \int_{\chi_{or}^2}^{+\infty} \wp(\chi^2, \nu)d\chi^2 \qquad (4.84)$$

where $\wp(\chi^2, \nu)$ is given by 4.83.

The result of this integral gives the total probability *in the universe* of obtaining a $\chi^2 \geq \chi_o^2$. The integral representing the total probability between $-\infty$ and $+\infty$ is $P(\chi^2, \nu)$, equal 1. The information of the value of $P(\chi^2, \nu)$ between $-\infty$ and χ_o^2 is the same as the one we have for $P(\chi^2, \nu)$ between χ_{or}^2 and $+\infty$, since one is the complement of the other, since the sum of $P(\chi^2, \nu)$ between $-\infty$ and χ_o^2 and of $P(\chi^2, \nu)$ between χ_o^2 and $+\infty$ gives exactly 1. Therefore, the larger is $P(\chi^2, \nu)$ between χ_o^2 and $+\infty$, the smaller is $P(\chi^2, \nu)$ between $-\infty$ and χ_o^2 and vice versa.

The example in figure 4.10, indicates the area below the curve for $P(\chi^2, \nu)$ between χ_o^2 and $+\infty$. Usually, these are exactly the values reported in the tables. These are values for which the result of the operation of the integral for the distribution function of χ^2, namely $P(\chi^2, \nu)$ expressed as a percentage or in centesimal values, for different values of the degrees of freedom, in the interval $\chi^2 \geq \chi_o^2$ or $\chi_r^2 \geq \chi_{ro}^2$, namely the $P(\chi_r^2 \geq \chi_{ro}^2)$, where χ_{ro}^2 is χ_o^2/ν. Otherwise, we can find the values of χ^2 corresponding to certain values of $P(\chi^2, \nu)$ or those for the χ_r^2, in the tables in the two cases. The table most directly useful is the one obtained in the following exercise and reported below as a result, expressed as a percentage for the values of $P(\chi_r^2 \geq \chi_{ro}^2, \nu)$, as a function of χ_r^2.

Exercise: *programme for the χ^2 tables*

Write a computer programme in an elementary language to:

calculate the values of the distribution function of χ^2 and print them in the interval between χ^2 0 and 50, of ν between 0 and 50;

calculate the values of the integral for the same values;

calculate the values of the integral for the same values, for $\chi^2 \geq \chi^2_{or}$, namely the $P(\chi^2 \geq \chi^2_{or}, \nu)$;

repeat for the distribution function of the χ^2_r, $\chi^2_{r_o}$ and the $P(\chi^2_r \geq \chi^2_{r_o}, \nu)$.

χ^2_r	$P(\chi^2_r)$	ν	χ^2_r	$P(\chi^2_r)$	ν	χ^2_r	$P(\chi^2_r)$	ν
0.001	0.99	1	0.111	0.99	5	0.256	0.99	10
0.004	0.95	1	0.229	0.95	5	0.394	0.95	10
0.016	0.90	1	0.322	0.90	5	0.487	0.90	10
0.064	0.80	1	0.469	0.80	5	0.618	0.80	10
0.148	0.70	1	0.600	0.70	5	0.727	0.70	10
0.455	0.50	1	0.870	0.50	5	0.934	0.50	10
0.708	0.40	1	1.026	0.40	5	1.047	0.40	10
1.074	0.30	1	1.213	0.30	5	1.178	0.30	10
1.642	0.20	1	1.458	0.20	5	1.344	0.20	10
2.706	0.10	1	1.847	0.10	5	1.599	0.10	10
3.841	0.05	1	2.214	0.05	5	1.831	0.05	10
6.635	0.01	1	3.017	0.01	5	2.321	0.01	10

Table 4.2: Values of probability $P(\chi^2_r \geq \chi^2_{r_o}, \nu)$, of the corresponding degrees of freedom ν and of the parameter χ^2_r. Example of the solving of the exercise for χ^2_r, with $\nu = 1$, $\nu = 5$ and $\nu = 10$ and some values of $P(\chi^2_r, \nu)$.

Coming back to what we said about $P(\chi^2_r \geq \chi^2_{r_o}, \nu)$, we suggest extensive use of this table for the evaluation of the hypothesis as well as for the experimental situation.

For values of $P(\chi_r^2 \geq \chi_{r_o}^2, \nu) \leq 50\%$ it is reasonable to ascribe greater validity to the hypothesis and to the measurements, with respect to the values for $P(\chi_r^2 \geq \chi_{r_o}^2, \nu) \geq 50\%$. On the other hand, it is impossible to establish valid a priori criteria for all hypotheses and experimental situations. It is clear that if the measurements are particularly accurate and the data are in relevant number, we expect a value of $P(\chi_r^2 \geq \chi_{r_o}^2, \nu) \ll 50\%$, even more if the hypothesis can be formulated with a simple analytical function. If, on the other hand, mea-

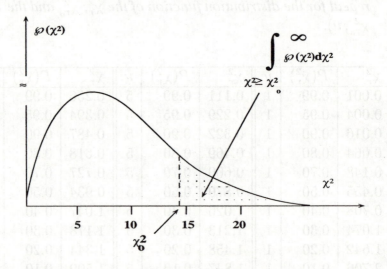

Figure 4.10: Curve of the probability distribution function, $\wp(\chi^2, \nu)$, for $\nu = 10$; also indicated is the area below the integral equal to $P(\chi^2 \geq \chi_{or}^2, \nu)$.

surements are few and/or scarcely accurate, we can expect to find a $P(\chi_r^2 \geq \chi_{r_o}^2, \nu) \geq 50\%$, and even more if it is not possible to find, on the basis of simple considerations (even on the plot which represents the data), a reasonable analytical function that describes them.

It is useful to underline once more that the χ^2 test verifies the validity of a hypothesis by verifying the differences $y_i^{experimental} - y_i^{theoretical}$

(namely the $y_i - a - bx_i$ of paragraph 4.4) and the relevance of the $s_i^{experimental}$, also verifying how far these are from the $\sigma_i^{theoretical}$. In fact it evaluates how large the $s_i^{experimental}$ are (or, in the case of a unique and direct measurement, how large the Δ_i are). This also translates into the evaluation of the width of the experimental distributions. If it is possible to make good predictions on the results expected from the experiment, the χ^2 test verifies the validity of the experimental approach, thus allowing us to evaluate whether or not the measurements are made with an accuracy comparable to that with which the predictions have been made.

However, if the experimental situation is not completely under control, or the hypotheses for the theoretical description of the experiment are difficult, then, instead of being helpful, the χ^2 value can *confuse* the information, giving the ratio between the knowledge on the first and the second and thus not allowing us to disentangle the two single contributions. For this reason we often use alternative tests which are described below.

Example: relation between the parameters of the Lorentz function, the Cauchy function, the Breit-Wigner function and the Gauss function in the magnetic induction experiment

We have already seen the case of the approximation of an analytical function of the Gauss type, in the case of the photoresistance experiment (the correspondence between the experimental curve and the function was purely formal and had no interpretation as a physical law.).

There are cases in which an analytical function best approximates a *bell* type behaviour similar to the Gauss one.

Let us consider the magnetic induction experiment mentioned previously (appendix refMagnetic induction between two solenoids). For certain values of frequency ω_p of the primary voltage, the secondary circuit of voltage, V_s, is resonant with the one of the primary, V_p. The

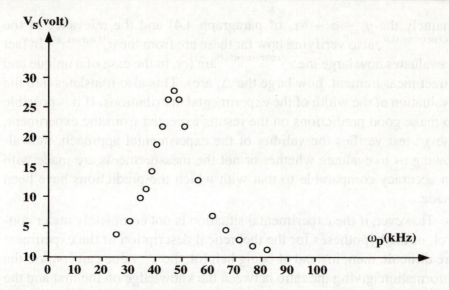

Figure 4.11: Curve of resonance for two coils in alternating voltage. The plot shows the voltage of the secondary circuit V_s, as a function of frequency ω_p of voltage V_p of the primary circuit. The curve is of the *Lorentz* or *Cauchy* or *Breit-Wigner* type and not of the Gauss type.

corresponding values of the frequencies are different for the different solenoids which are the secondary circuits, depending on the number of windings. The curve obtained is similar to the one in figure 4.11.

The curve is slightly asymmetric around the maximum value. There are other analytical functions (which are also distribution functions) which better approximate the curve with respect to the function of the Gauss type.

The function

$$f(x, \mu, \Gamma) = \frac{1}{\pi} \frac{\Gamma/2}{(x - \mu)^2 + (\Gamma/2)^2} \tag{4.85}$$

is known as the *Lorentz probability distribution function* .

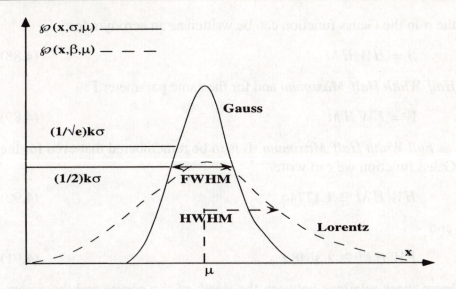

Figure 4.12: Comparison of two functions of Gauss and Lorentz (or *Breit-Wigner*) with the same mean value μ.

It is a special case of a function of the type

$$f(x) = \frac{1}{\pi} \frac{1}{1 + x^2} \qquad (4.86)$$

known as *the Cauchy function*. It is to be noted that the Lorentz distribution function was not introduced in the previous chapter because it cannot be directly derived from the Binomial distribution function like the other functions obtained as an approximation from it.

Equation 4.85 is similar to the Gauss function also in its parameters. In fact, it is :

$$f(x, \mu, \beta) = \frac{1}{\pi} \frac{\beta}{(x - \mu)^2 + (\beta)^2} \qquad (4.87)$$

with β being the Half Width Half Maximum. The parameter similar to the σ in the Gauss function can be written as an acronym for:

$$\beta = HWHM \tag{4.88}$$

Half Width Half Maximum and for the same parameter Γ:

$$\Gamma = FWHM \tag{4.89}$$

as *Full Width Half Maximum*. It is to be remembered that even for the Gauss function we can write

$$HWHM \cong 1.1774\sigma \tag{4.90}$$

and

$$FWHM \cong 2.3548\sigma \tag{4.91}$$

From these relations between the width of the curves and the parameters of the distribution, we can see that, for the Lorentzian one, the dependence is inversely proportional to the square of β by increasing x instead of depending from the inverse of the exponential to the square of σ [36]. That is, the curve, as seen also in figure 4.12, decreases less rapidly [37].

On making the approximations in the applications, the Lorentz function has to be treated in the same way as the Gauss function. It has two parameters very similar to it (see the discussion on least squares for the Gauss function).

---

[36] Analytically the two functions, although apparently quite similar, have a substantial difference, at the basis of which is the reason why the Cauchy function was not introduced among the functions of distribution discussed before. The standard deviation, which is the true σ, not only in the analytical sense but also as a parameter of the parent distribution probability function cannot be defined because the integral that rigorously defines standard deviation, which we have introduced as the limit of the sum, gives an ever increasing numerical result for x going to ∞.

[37] The graphic behaviour of the resonance function is common to several physical phenomena which recall resonance as the distribution of the spectra lines of emission and absorption. The parameters are different for different processes.

4.6 The t accuracy test

The χ^2 test has certain limitations. First of all, its use requires the formulation of a hypothesis with an adequate parent distribution function. Furthermore, in the definition appear both the standard deviations of the parent distribution function and that of the data.

It is possible to evaluate the quality of measurements also by comparing the two samples of similar measurements, instead of with an analytical function given by a theoretical hypothesis. This procedure is used especially for quantitative evaluations on variables not corresponding to physical quantities. It is widely used especially in biomedical disciplines.

It is possible to make tests based on the observed variables (for example, the standard deviation or the weighted mean) of the experimental distribution. The parameters are defined as functions of such variables.

To use these tests, it is necessary, as will be done below, to discuss the criteria for acceptability, that is, as we have seen in the example of the χ^2, for which probability values we can evaluate the accuracy of the test and the hypothesis. This is the reason why we will again discuss the parameters of the test at the end of the chapter and discuss them in the context of probability intervals.

We now introduce other tests, which are all based on the variables of parent distributions.

The test we are going to introduce verifies the hypothesis that series of measurements are e.g. compatible or uncompatible.

Let us consider a simple case with only two distributions of data, the x_i, with experimental standard deviations s_i^x, with arithmetical mean $\overline{x}$ and mean experimental standard deviation, $s_{\overline{x}}$. We wish to compare this data sample with a similar one, the measurements z_i, with experimental standard deviations s_i^z, with arithmetical mean $\overline{z}$ and mean experimental standard deviation, $s_{\overline{z}}$.

The two samples of measurements can be composed of a different number of data.

The arithmetical mean alone cannot provide an adequate estimate. It is not enough to verify that within certain intervals to be defined, the arithmetical means of the two distributions are comparable and therefore to conclude that one is a more reliable series of measurements than the other. It is necessary to refer each series to its own characteristics. A series can be composed of data collected with greater accuracy than another one. Therefore it can be more or less compatible with another one within the experimental standard deviation. To quantify this concept, we use a parameter that is equal to the arithmetical mean with respect to its own mean experimental standard deviation:

$$t_x = \frac{\overline{x}}{s_{\overline{x}}} \tag{4.92}$$

This is a mean referred to the deviation, to be compared, for example, with:

$$t_z = \frac{\overline{z}}{s_{\overline{z}}} \tag{4.93}$$

of an homologous distribution.

The variable t_x or t_z or simply t, given by the ratio of the arithmetical mean and the experimental standard deviation, is very similar to the one commonly called *Student's* t [38] which will be introduced formally in section 4.9.1. For simplicity, we will call it here t parameter. In practical terms it is:

$$t = \frac{\overline{x} - \mu}{s_{\overline{x}}} \tag{4.94}$$

when $\mu = 0$.

---

[38]Mr *Student* never existed. It is customary to write *Student* with a capital letter because this is the pseudonym used by the mathematician Gosset who published the article introducing for the first time a parameter similar to the one here called t.

Let us now see how the parameter t can be used, similarly to χ^2, to evaluate the accuracy of a series of measurements.

Let us first of all consider the special case of a series of measurements x_i, with s_i^x constant, namely:

$$s_i^x \equiv s_{i-1}^x \equiv s_{i+1}^x \equiv s_x \tag{4.95}$$

Let us also suppose that the number of degrees of freedom is given by $\nu = n - 1$. Rewriting the definition:

$$t_x = \frac{\overline{x}}{s_{\overline{x}}} = \frac{\overline{x}}{s_{\overline{x}}} \frac{\sqrt{n}}{\sqrt{n}} \tag{4.96}$$

having multiplied and divided by $\sqrt{n}$. From the definition of $s_{\overline{x}}$, and from equation 4.95:

$$s_{\overline{x}} = \frac{s_x}{\sqrt{n}} \tag{4.97}$$

namely

$$s_x = s_{\overline{x}}\sqrt{n} \tag{4.98}$$

Substituting this expression in equation 4.96

$$t_x = \frac{\overline{x}\sqrt{n}}{s_{\overline{x}}\sqrt{n}} = \frac{\overline{x}\sqrt{n}}{s_x} = \frac{\overline{x}\sqrt{n}}{s_x} \frac{\sqrt{n-1}}{\sqrt{n-1}} \tag{4.99}$$

having multiplied and divided by $\sqrt{n-1}$, the square root of the degrees of freedom, $\sqrt{n-1} = \sqrt{\nu}$. Therefore $s_x = constant \equiv s$, we can rewrite 4.99:

$$t_x \equiv t = \frac{\overline{x}\sqrt{n}}{s} \frac{\sqrt{\nu}}{\sqrt{\nu}} \tag{4.100}$$

Recalling equation 4.71 of paragraph 4.5.2, we have:

$$\frac{s^2}{\sigma^2} \cong \frac{\chi^2}{\nu} \tag{4.101}$$

namely :

$$\frac{s}{\sigma} \cong \frac{\sqrt{\chi^2}}{\sqrt{\nu}} \tag{4.102}$$

that is

$$s\sqrt{\nu} \cong \sigma\sqrt{\chi^2} \tag{4.103}$$

Therefore equation 4.100, can be written as:

$$t \cong \frac{\overline{x}}{\sigma\sqrt{\chi^2}}\sqrt{n}\sqrt{\nu} \tag{4.104}$$

Since, for similar approximations, $\sigma_j = constant = \sigma$ are the standard deviations of the parent distributions of each single experimental value x_j, obtained as a mean, in equation 3.106, we can write:

$$\overline{\sigma} = \frac{\sigma}{\sqrt{n}} \tag{4.105}$$

We can therefore see the relation between χ_r^2 and the parameter t:

$$t \cong \frac{\overline{x}}{\sigma_\mu}\sqrt{\frac{1}{\chi_r^2}} \tag{4.106}$$

Therefore the parameter t depends upon the degrees of freedom and is similar to parameter χ^2.

   To use the parameter rigorously as a test, we introduce, similarly to what we did for χ^2, the probability distribution function for the t parameter, given a certain number of degrees of freedom ν. This is :

$$\wp(t, \nu) = \frac{\Gamma(\frac{\nu+1}{2})}{\sqrt{\pi\nu}\Gamma(\frac{\nu}{2})}(1 + \frac{t^2}{\nu})^{-\frac{\nu+1}{2}} \tag{4.107}$$

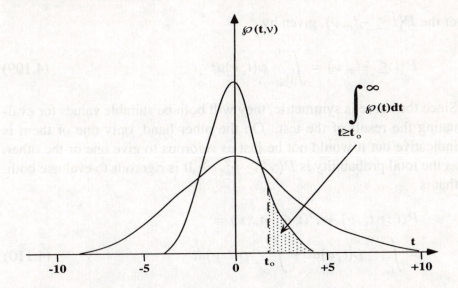

Figure 4.13: The curve of the probability distribution function, $\wp(t, \nu)$, for $\nu = 1$ and $\nu = 10$. The value of $\wp(t, 1) = 0.5$ is indicated in the ordinate as an example of $t = 0$. As an example of use a part of the area for the curves $\nu = 10$ is indicated for a certain t_o (see definition in equation 4.94).

The use of parameter t is therefore based on the values that the distribution function assumes for a certain number of degrees of freedom.

Figure 4.13 shows the curve for $\nu = 1$ and $\nu = 10$. The curve is symmetric and becomes increasingly narrower the higher the number of degrees of freedom.

Also for $\wp(t, \nu)$, the test will give a positive result and will be more reliable the lower its total probability, the $P(t \geq t_o, \nu)$, given by

$$P(t \geq t_o, \nu) = \int_{t_o}^{+\infty} \wp(t, \nu)dt \qquad (4.108)$$

or the $P(t \leq -t_o, \nu)$, given by

$$P(t \leq -t_o, \nu) = \int_{-\infty}^{-t_o} \wp(t, \nu)dt \tag{4.109}$$

Since the curve is symmetric, they will both be suitable values for evaluating the result of the test. On the other hand, only one of them is indicative but it would not be just as rigorous to give one or the other, as the total probability is $P(\chi^2 \geq \chi_o^2, \nu)$. It is rigorous to evaluate both, that is

$$P(t \geq t_o, \nu) + P(t \leq -t_o, \nu) =$$
$$= \int_{t_o}^{+\infty} \wp(t, \nu)dt + \int_{-\infty}^{-t_o} \wp(t, \nu)dt \tag{4.110}$$

Since the curve is symmetric,

$$P(t \geq t_o, \nu) = P(t \leq -t_o, \nu) \tag{4.111}$$

then

$$P(t \geq t_o, \nu) + P(t \leq -t_o, \nu) = 2 \int_{t_o}^{+\infty} \wp(t, \nu)dt \tag{4.112}$$

To avoid ambiguities it is customary to tabulate and evaluate on the basis of the <u>total</u> probability value from equation 4.112.

Let us examine an example of application of parameter t for hypothesis testing.

Example: Student's test for the difference between two means.

Let us consider two ensembles of measurements x_i and z_i, in numbers n_x and n_z.

We wish to know how plausible it is that $\overline{x} = \overline{z}$, namely that the arithmetic means are equal.

This is tantamount to asking how plausible it is that $\overline{x} - \overline{z} = 0$. Let us see how the test of parameter t can help us to evaluate it.

We wish to know the probability of $\overline{x} - \overline{z} = 0$. We translate this into a probability of parameter t for evaluation with equation 4.112.

Given the s_x, s_z, the $s_{\overline{x}}$ and the $s_{\overline{z}}$, we define $d = \overline{x} - \overline{z}$.

It is possible to apply the parameter t test, now defined as

$$t_d = \frac{d}{s_d} \tag{4.113}$$

with s_d, the experimental standard deviation obtained from the combination of the experimental standard deviations $\overline{s}_x$ and $\overline{s}_z$, namely

$$s_d^2 = s_{\overline{x}}^2 + s_{\overline{z}}^2 = \frac{s_x^2}{n_x} + \frac{s_z^2}{n_z} \tag{4.114}$$

Let us suppose that $s_x = s_z \equiv s$, to make interpretation of the result and the discussions that follow more immediate. This hypothesis is not a limiting one; even the parent distributions must have $\sigma_x = \sigma_z = \sigma$. With this hypothesis equation 4.114 becomes

$$s_d^2 = \frac{s^2}{n_x} + \frac{s^2}{n_z} = (\frac{1}{n_x} + \frac{1}{n_z})s^2 \tag{4.115}$$

and correspondingly

$$\sigma_d^2 = (\frac{1}{n_x} + \frac{1}{n_z})\sigma^2 \tag{4.116}$$

Coming back the expression of χ^2 4.72, we obtain respectively, for each distribution:

$$\chi_x^2 \cong \frac{(n_x - 1)s_x^2}{\sigma^2} \tag{4.117}$$

$$\chi_z^2 \cong \frac{(n_z - 1)s_z^2}{\sigma^2} \tag{4.118}$$

For the evaluation as a test on d, we must obtain the combined distribution of χ^2 for the two series. This is possible due to the fact that the measurements by which we can define χ^2 are independent, and parameters χ^2 can be added , because we can combine the probability distributions of χ^2. In this way (with a total number of degrees of freedom equal to $\nu = n_x - 1 + n_z - 1$), we can write:

$$\chi^2 = \chi_x^2 + \chi_z^2 \cong$$

$$\cong \frac{1}{\sigma^2}((n_x - 1)s_x^2 + (n_z - 1)s_z^2) =$$

$$= \frac{1}{\sigma^2}(n_x s^2 + n_z s^2 - 2s^2) =$$

$$= \frac{1}{\sigma^2}(n_x + n_z - 2)s^2 =$$

$$= \frac{1}{\sigma^2}\nu s^2 \qquad\qquad (4.119)$$

Therefore:

$$s^2 \cong \frac{\sigma^2 \chi^2}{\nu} \qquad\qquad (4.120)$$

Coming back to equation 4.115 we have:

$$s_d^2 = (\frac{1}{n_x} - \frac{1}{n_z})s^2 \cong (\frac{1}{n_x} - \frac{1}{n_z})\frac{\sigma^2 \chi^2}{\nu} \qquad\qquad (4.121)$$

and correspondingly

$$\sigma_d^2 = (\frac{1}{n_x} - \frac{1}{n_z})\sigma^2 \cong (\frac{1}{n_x} - \frac{1}{n_z})\frac{\nu s^2}{\chi^2} \qquad\qquad (4.122)$$

Therefore

$$\frac{1}{n_x} - \frac{1}{n_z} \cong \frac{\sigma_d^2}{\frac{\nu s^2}{\chi^2}} \qquad\qquad (4.123)$$

on substituting equation 4.121, in equation 4.123, we have

$$s_d^2 \cong \frac{\sigma_d^2}{\frac{\nu s^2}{\chi^2}} \frac{\sigma^2 \chi^2}{\nu} \tag{4.124}$$

But

$$s_d^2 \cong \frac{\sigma^2 \chi^2}{\nu} \tag{4.125}$$

thus

$$s_d^2 \cong \frac{\sigma_d^2 \chi^2 \chi^2 \sigma^2}{\frac{\nu^2 s^2 \chi^2}{\nu}} = \frac{\sigma_d^2 \chi^2}{\nu} =$$

$$= \frac{\sigma_d^2 \chi^2}{s^2} \cdot \frac{\sigma^2 \chi^2}{\nu} = \sigma_d^2 \chi^2 \tag{4.126}$$

Therefore we can define the parameter t_d by analogy:

$$t_d = \frac{\overline{x} - \overline{z}}{s_d} \cong \frac{d}{\sqrt{\sigma_d^2 \chi^2}} \tag{4.127}$$

At this point, we can evaluate the probability that $\overline{x} - \overline{z} = 0$, as we wished to verify. This is tantamount to asking about the probability that $t_d = 0$. We are therefore asking how much is $P(t \geq t_d, \nu)$. If it is small, the hypothesis or the statement that $\overline{x} - \overline{z} = 0$ is plausible. This would in fact mean that the integral, as the area below values t_d at $+\infty$, is small. Therefore it is not probable that another t_d can be found by taking another series of measurements with which to compare x_i or z_i and their respective $\overline{x}$ or $\overline{z}$.

It is worth underlining that the tables often give the values of t and of ν, for which $P(t, \nu) = 0.05$ or $P(t, \nu) = 0.01$. That is to say, the tables do not give possible results of the integral corresponding to ν, as in the case of χ^2, but rather the values ν and t for which the probability is 5% or 1%.

For practical use with experimental distributions, it is best to rewrite the expression of t_d in equation 4.127

$$t_d = \frac{\overline{x} - \overline{z}}{\sqrt{\frac{s_x^2}{n_x} + \frac{s_z^2}{n_z}}} \cong \frac{\overline{x} - \overline{z}}{\sqrt{\sigma_d^2 \chi^2}} \tag{4.128}$$

since, from equation 4.122,

$$\sigma_d^2 = (\frac{1}{n_x} - \frac{1}{n_z})\frac{\nu s^2}{\chi^2} \tag{4.129}$$

we have

$$t_d \cong \frac{\overline{x} - \overline{z}}{\sqrt{(\frac{1}{n_x} - \frac{1}{n_z})\nu s^2}} = \frac{\overline{x} - \overline{z}}{\sqrt{(\frac{1}{n_x} - \frac{1}{n_z})(n_x + n_z - 2)s^2}} \tag{4.130}$$

It must be kept in mind that in the last expression factor $n_x - n_z - 2$ is due to the fact that we limited ourselves to the simple case of $\nu = n - 1$ degrees of freedom for each series.

n	$Clemens(C)$	$Ulrich(U)$
1	1°06'00"	0°12'00"
2	2°00'30"	1°09'00"
3	0°40'00"	2°12'00"
4	1°30'00"	3°12'00"
5	0°50'30"	2°13'30"

Table 4.3: Experimental measurements detected by two students for the value of 0 of a polarimeter.

Exercise: *Student's t test with the ruler*

Let us consider two series of independent measurements of lengths of an edge of this sheet made with a ruler with resolution $s \equiv \Delta l = 0.05cm$; one series for 7 data, the other for 61:

- *determine t_d;*

- *look for the probability for the result obtained in the tables;*

- *ascribe greater or lesser consistence to the two series of measurements.*

Example: The three-fold case of measurements with the polarimeter

Let us consider four series of independent measurements made by two groups of students. They wish to measure the position of equal-shadow of the polarimeter, in line with what we discussed previously. Suppose that two experimentalists in the same conditions have obtained the two series of measurements shown in the table 4.3 (Ulrich and Clemens) and the other two (Teo and Jessica), those in table 4.4.

n	Teo	Jessica	n	Teo	Jessica	n	Teo	Jessica
1	4°36'00"	2°03'30"	41	3°40'00"	4°39'30"	81	3°03'00"	3°10'00"
2	4°39'00"	7°01'00"	42	1°20'30"	1°01'30"	82	4°20'00"	4°31'30"
3	3°43'30"	5°30'00"	43	0°45'30"	0°34'00"	83	2°03'00"	2°10'30"
4	3°33'00"	5°06'00"	44	1°01'00"	1°03'30"	84	1°01'30"	1°05'30"
5	5°07'30"	6°03'00"	45	2°10'30"	3°15'00"	85	6°40'30"	6°33'00"
6	3°03'00"	3°06'00"	46	1°10'30"	1°03'00"	86	4°05'30"	5°06'00"
7	3°04'30"	4°06'00"	47	3°24'00"	3°10'00"	87	3°09'00"	4°12'00"
8	2°42'00"	2°13'30"	48	1°10'00"	1°03'00"	88	2°09'30"	1°01'00"
9	5°04'30"	5°10'30"	49	3°07'30"	3°10'00"	89	0°11'30"	1°20'30"
10	0°42'00"	0°45'30"	50	3°07'00"	3°09'00"	90	2°33'30"	2°40'30"
11	4°01'30"	4°36'00"	51	4°03'30"	4°01'30"	91	2°46'30"	2°50'00"
12	3°43'30"	4°13'30"	52	1°03'00"	1°03'00"	92	1°20'00"	1°13'30"
13	3°36'00"	4°12'30"	53	4°30'00"	4°42'00"	93	1°12'00"	2°20'00"
14	4°06'00"	4°07'30"	54	0°21'00"	0°10'30"	94	3°03'00"	4°20'00"
15	2°01'30"	2°13'30"	55	0°46'30"	0°12'30"	95	2°10'30"	2°12'30"
16	1°42'00"	4°07'30"	56	4°10'30"	4°37'30"	96	0°09'00"	0°10'30"
17	2°31'30"	3°05'30"	57	6°31'30"	6°10'00"	97	3°03'00"	4°20'00"
18	0°52'30"	0°28'30"	58	2°10'00"	1°40'30"	98	1°07'30"	1°05'30"
19	3°31'30"	4°07'30"	59	1°01'30"	1°03'00"	99	2°03'00"	2°20'00"
20	2°06'00"	2°06'00"	60	3°30'00"	3°40'30"	100	1°11'30"	2°20'00"
21	1°21'00"	1°12'00"	61	1°02'00"	1°33'00"	101	3°10'30"	3°12'00"
22	1°54'00"	1°03'00"	62	1°32'00"	1°30'00"	102	2°13'30"	2°10'00"
23	0°40'30"	0°01'30"	63	2°12'00"	3°20'00"	103	0°02'30"	0°40'30"
24	4°01'30"	4°01'30"	64	3°06'00"	3°07'00"	104	4°40'30"	4°52'30"
25	1°04'30"	1°05'30"	65	4°24'00"	5°30'00"	105	2°10'30"	2°09'00"
26	4°33'00"	4°01'30"	66	3°13'30"	3°14'30"	106	3°12'00"	3°36'00"
27	3°10'00"	3°03'00"	67	2°02'30"	2°01'30"	107	4°27'00"	4°25'30"
28	3°19'30"	3°15'30"	68	1°01'30"	1°12'00"	108	0°31'30"	1°02'00"
29	3°33'00"	3°40'00"	69	1°03'00"	1°10'30"	109	2°11'30"	2°09'00"
30	3°40'00"	3°03'00"	70	1°01'30"	1°01'30"	110	5°38'00"	5°21'30"
31	3°58'30"	1°01'30"	71	1°10'00"	1°31'30"	111	3°20'00"	3°31'30"
32	5°03'00"	5°01'30"	72	1°05'00"	1°03'00"	112	5°07'30"	5°11'00"
33	3°31'30"	3°07'30"	73	6°09'30"	5°03'30"	113	2°24'00"	2°11'30"
34	0°54'00"	1°04'30"	74	4°01'30"	3°10'00"	114	1°12'00"	2°20'00"
35	1°04'30"	2°03'30"	26	4°33'00"	4°01'30"			
36	1°03'30"	1°04'30"	76	0°48'00"	1°30'00"			
37	1°05'00"	1°06'30"	77	2°31'30"	2°20'00"			
38	1°16'30"	2°07'00"	78	2°05'30"	2°04'30"			
39	1°12'00"	1°04'30"	79	2°39'00"	2°10'00"			
40	3°07'00"	3°06'00"	80	0°10'30"	0°01'30"			

Table 4.4: Experimental measurements detected by two students for the value of 0 of the polarimeter.

It is possible to determine the corresponding t, with $n = 5$ and $\nu = 4$. For the first couple of students we obtain:

$$\overline{C} = 1°13'30" \qquad \overline{U} = 2°13'30" \tag{4.131}$$

$$s_C = 0°32'30" \qquad s_U = 1°47'00" \tag{4.132}$$

$$s_{\overline{C}} = 0°15'00" \qquad s_{\overline{U}} = 1°30'30" \tag{4.133}$$

from which

$$t_C \cong 5.0 \qquad t_U \cong 3.5 \qquad (4.134)$$
$$P(t \geq t_C) \cong 99.5\% \qquad P(t \geq t_U) \cong 98.0\% \qquad (4.135)$$

The test on the single series of measurements is not very significant. It simply gives a measurement of data dispersion around the mean value if this is close to zero. The parameters s_C, s_U and $s_{\overline{C}}$, $s_{\overline{U}}$ alone give the same information. If the value of t corresponds, as in the special case for Clemens's series of measurements, to a very high value of $P(t \geq t_o)$, we are simply verifying that the data distribution is Gauss and therefore that there are no data on the queues of the distribution curves of variable t, with mean value 0.

Therefore, applying the t test to the single series of measurements does not give much information. The only thing we can conclude from the comparison of the two probability values is that the measurements are reasonably distributed without too high a dispersion, which is in fact comparable with the indicated instrumental resolution of $0°, 01', 30"$, even though for one series the dispersion is twice that of the other. We should thus be tempted, on the basis of $s_{\overline{C}}$ and $s_{\overline{U}}$ to infer that Ulrich is not as careful an experimentalist as Clemens in using the instrument.

To compare the two series of measurements we can also use the t test in another way.

By defining, as in the previous paragraph, the difference $\overline{U} - \overline{C}$, then

$$s_d = \sqrt{\frac{s_U^2}{5} + \frac{s_C^2}{5}} = 0°34'00" \qquad (4.136)$$

$$t_d = \frac{\overline{U} - \overline{C}}{s_d} = 0.99902 \cong 1.0 \qquad (4.137)$$

From the tables, now having $n = 10$ and $\nu = 8$:

$$P(t \geq t_d) \cong 65\% \qquad (4.138)$$

From these values we can conclude that the two series of measurements are plausible and relatively compatible, even though their discrepancy has a quite high (about 35%) probability of occurring. But then what we have concluded before on the experimentalist M is weaker. If the compatibility between the two measurements is established in this way, it is not legitimate to infer that one is better than the other. However, we do not have any valid reason for believing more in this statement instead than in the previous one. The probability values should be a suitable parameter in discriminating. We could even judge both the probability values obtained in the two cases compatible with the opposite conclusions. Continuing this reasoning, nothing prevents us from stating that these are equal. Or better, we cannot decide *within what limits* they can be considered equal or different. In this case it would be even more difficult to discriminate between the two conclusions. The tables themselves are often misleading, especially those presenting the probability values between 1% and 5%. Without suitable premises, these values are generic and do not contribute any decisive elements.

All this leads us to the need to discuss the probability and intervals of probability values that give a certain confidence in the result. This issue will be discussed in the following paragraph.

An alternative route is that of making other tests and then discussing the results again. This procedure is quite similar to what we have done in this case, by evaluating two different presentations of parameter t. Another test is for example the $\mathcal{F}$ test, whose results on the same series of measurements are given in the following paragraph.

For completeness, let us now advance the hypothesis, as in the previous chapter, that $s_U \cong s_C \equiv s = 0°50'00"$. Then we can define

$$s_d = \sqrt{\frac{s_C^2}{5} + \frac{s_U^2}{5}} = \sqrt{\frac{2s^2}{5}} = 0°32'00" \tag{4.139}$$

and

$$t_d = \frac{\overline{U} - \overline{C}}{s_d} = 1.061 \cong 1.1 \tag{4.140}$$

From the tables we find

$$P(t \geq t_d) \cong 65\% \tag{4.141}$$

Therefore we obtain a conclusion similar to the previous estimate of t_d.

The student can repeat the exercise for table 4.4. To be noted in particular is the number of measurements at which the t test loses significance, with the distribution becoming increasingly more of Gauss type with the number of the measurements.

4.7 The $\mathcal{F}$ accuracy test

The $\mathcal{F}$ parameter test is similar to the t parameter test.

Instead of evaluating the compatibility between two series of measurements, on the basis of the arithmetical mean of the measurements, $\mathcal{F}$ evaluates the discrepancy between experimental standard deviations. If, for example, we wish to see if the experimental standard deviations s_x and s_z of two independent series of measurements x_i and z_i are equal, and how plausible this statement is, we can evaluate the probability distribution function for the parameter $\mathcal{F}$ defined as

$$\mathcal{F} = \frac{s_x^2}{s_z^2} \tag{4.142}$$

which can also be written as

$$\mathcal{F} \cong \frac{\frac{\chi_x^2}{\nu_x} \sigma_x^2}{\frac{\chi_z^2}{\nu_z} \sigma_z^2} \tag{4.143}$$

Since the two series of measurements are independent, the corresponding probability function, on indicating with $\nu = \nu_x + \nu_z$, is :

$$\wp(\mathcal{F}, \nu_x, \nu_z) = \frac{\left(\frac{\nu_x}{\nu_z}\right)^{\frac{\nu_x}{2}} \Gamma\left(\frac{\nu}{2}\right)}{\Gamma\left(\frac{\nu_x}{2}\right)\Gamma\left(\frac{\nu_z}{2}\right)} \left(\frac{\nu_x}{\nu_z}\mathcal{F} + 1\right)^{-\frac{\nu}{2}} \mathcal{F}^{\frac{\nu_x-2}{2}} \tag{4.144}$$

The curves of the probability function $\wp(\mathcal{F}, \nu_x, \nu_z)$ are represented in figure 4.14. As we can see, the curves do not change much, even though the number of degrees of freedom varies.

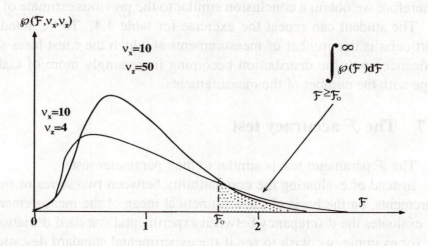

Figure 4.14: Curve of the probability distribution function, $\wp(\mathcal{F}, \nu_x, \nu_z)$, for two pairs of values of degrees of freedom $\nu_x = 10, \nu_z = 50$ and $\nu_x = 10, \nu_z = 4$. The maximum value of $\wp(\mathcal{F}, 10, 50)$ tends to $\wp(\mathcal{F}, \nu_x, \nu_z) = 1$ when $\mathcal{F} = 1$. The maximum value for $\wp(\mathcal{F}, 10, 4)$ is instead not much higher than 0.5, but corresponds to the values of $\mathcal{F} < 1$. As a practical example, a part of the area for curves $\nu_x = 10$, $\nu_z = 50$ $\nu_x = 10$, $\nu_z = 4$ and for a certain $\mathcal{F}_o$ is indicated. It is to be noted that the area below is, for large values of $\mathcal{F}_o$, almost similar in the two cases.

It is necessary to evaluate the total probability in this case as well.

$$P(\mathcal{F} \geq \mathcal{F}_o, \nu_x, \nu_z) = \int_{\mathcal{F}_o}^{+\infty} \wp(\mathcal{F}, \nu_x, \nu_z)d\mathcal{F} \qquad (4.145)$$

As can be seen in figure 4.14, even the values of $P(\mathcal{F} \geq \mathcal{F}_o, \nu_x, \nu_z)$ do not vary greatly with the variation of ν. This is the reason why we very often find in the tables the corresponding values of the degrees of

freedom in a restricted interval for certain values of given total probability. The contrary is always more useful. To guide the student it is
important to underline that the total probability values, defined at 5% or
1% in correspondence to which we can read the values of $\mathcal{F}$ and ν, are
those of most frequent use in practice. This is similar to the comment
made previously: the $\wp(\mathcal{F}, \nu_x, \nu_z)$ function has similar values of $\mathcal{F}$ even
for large differences in the degrees of freedom.

It is necessary to pay attention to the interpretation of the ratio that
defines $\mathcal{F}$, that is, s_x^2/s_z^2. What we wish to prove is if one standard
deviation is larger than the other and how reliable this statement is. To
apply the test itself, however, we should know the result a priori. In
fact, the tables with which probability is evaluated assume that s_x is
larger than s_z and as such they give the value of $\mathcal{F}$. This is only a
convention on which most tables are based. They show the ratios with
the standard deviation that appears numerically higher in the numerator.
In some treatments there is often a discussion on how convenient the
further verification of the reverse ratio $1/\mathcal{F}$ is. In reality, this is not
necessary but in any case it is easy to do. In fact, in the tables at fixed
values of the integral of probability, which, as stated above, are the most
common, we can find both values. If one of the two does not fall within
the tables, the test can be performed in any case, provided that suitable
caution is used in reading the correct degrees of freedom. If neither of
the two is within the tables, the test does not lead to any conclusion.

Exercise: $\mathcal{F}$ test with the ruler

*Let us consider two series of independent measurements of lengths
of one side of this sheet, made with a ruler with resolution $s_x \equiv
\Delta_x l = 0.05cm$, and with another one with resolution $s_z \equiv \Delta_z l =
0.1cm$, with the first series of 25 data, the other of 131:*

- *determine $\mathcal{F}$;*

- *look for the probability of the result obtained in the tables;*

*- ascribe a larger or smaller consistency to the two series of mea-
surements.*

Example: $\mathcal{F}$ for the standard deviations of the measurements with the polarimeter

Let us go back to values determined in the previous paragraph for the polarimeter experiment. We obtained the values

$$s_C = \quad 0°32'30''$$
$$s_U = \quad 1°09'30''$$

We can evaluate the consistency of the two series of measurements, by evaluating

$$\mathcal{F}_{CU} = \frac{s_C^2}{s_U^2} = 0.22 \tag{4.146}$$

or similarly

$$\mathcal{F}_{UC} = \frac{s_U^2}{s_C^2} = 4.6 \tag{4.147}$$

For $\nu_C = 4$ and $\nu_U = 4$, the corresponding probabilities are:

$$P(\mathcal{F} \quad \geq \mathcal{F}_{CU}) \gg 5\%$$
$$P(\mathcal{F} \quad \geq \mathcal{F}_{UC}) \cong 90\%$$

As we said, it is common practice always to define $\mathcal{F}$ with the larger standard deviation in the numerator, especially in the biomedical field, an application of which is described in the following paragraph.

This definition is justified in certain applications as the so-called *ANOVA*, described later on, in which s^2 in the denominator is a $s_{\bar{x}}^2$, a mean standard deviation, which we expect to be smaller or equal to s_x^2 in all cases.

In this example, there is no reason to expect a difference between s_U and s_C, but on the contrary there is a clear difference.

Therefore, we would again tend to conclude the same as from the value of the standard deviations, namely that Ulrich did not take the measurements with the same accuracy as Clemens. This would mean that he has probably made an illegitimate error, at least in several readings.

If this were the case, the value of t_d would lead us to the same conclusion, which we have seen is not justified.

Therefore, as we mentioned in chapter 1.2.1, there must be a systematic error, which adds up to the *offset* zero error of the scales (see figure 1.6).

From the comparison between $\mathcal{F}_{CU}$ and $\mathcal{F}_{UC}$, we can only conclude that since there are no evident reasons why $s_U > s_C$, or vice versa, and since we obtained two different probability values, the test indicates the presence of a systematic error, and therefore it does not lead to any quantitative conclusion. We will return to this point in the following paragraph.

This example is useful in demonstrating that it is not sufficient to obtain only one apparently *good* value of $P(\mathcal{F} \geq \mathcal{F}_o)$.

4.7.1 The $\mathcal{F}$ test for the analysis of one variance or ANOVA

The tests introduced in the previous examples with the parameter $\mathcal{F}$ are generally indicated as *analysis of variance* or *ANOVA* [39]. In different textbooks, especially those for biomedical applications , the test is frequently used. The reason for this is that in biomedical or biological experiments the data sample often contains data that vary among themselves, and are unclassifiable in one of the categories of uncertainty that we introduced in the first chapter because their effects are often con-

---

[39]It comes from the English acronym ANalysis Of VAriance or in the literature ambiguously also called ANalysis of One VAriance.

fused. This is not due to the fact that there are other types of uncertainties, which we have not introduced (or that are not deducible from them), but rather to the fact that it is not possible to investigate any further on a biological ensemble of data (*sample*). It is stated, in different texts, that there is a *biological variability* of data being analysed, and this must be taken into account. This statement can be corrected, because the organisms do present a diversification, but this is no different from that of any other measurement, whether of concentration, current, percentage of prime rate or any other measurement of which we always have limited knowledge.

$days$	$mice$
$j = 1, \cdots m$	$the = 1, \cdots n$
$m = 31$	$n = 21$
$\nu_g = 30$	$\nu_c = 20$
$s_{\overline{x}} = 0.6$	$s_{xj} = 1.24$
$$\mathcal{F} = \frac{s_{xj}^2}{s_{\overline{x}}^2}$$	

Table 4.5: Number of measurements, degrees of freedom and experimental standard deviations for the example of application of the ANOVA test.

If we have, for example, 21 mice on which to evaluate the quantity of liquids they can take in a day, we will find a distribution of the volume of the liquid which should be similar to a Gauss distribution, with a value of volume equal to the arithmetical mean. We can use a $\mathcal{F}$ (or ANOVA) test, comparing two series of measurements taken by the same experimentalist (hopefully as much as possible in the same physical-chemical conditions of temperature, pressure etc.) the day after. We have two s_x and s_y of the two days x and y. We define $\mathcal{F}_{xy}$ (if $s_x > s_y$) and we read in the tables the consistency as illustrated above. Alternatively we could define $\mathcal{F}_{\overline{xy}}$ from $s_{\overline{x}}$ and $s_{\overline{y}}$ but the parameter would be different only if the number of measurements were not the same. In this case it would not be advisable to use the test because it would give a biased value of $\mathcal{F}$, since a priori one of the two standard deviations of the mean would already be larger than the other. This must be taken as a warning in the use of variance tests.

There are cases which more commonly take the name of ANOVA, in which one has to compare the standard deviations to check two different types of deviations in the data. One of a series of data $j - th$ x_j, for example s_{xj} calculated for $i =, \cdots, n$ and the other of the means between the series of data, calculated for $j = 1, \cdots, m$ series of measurements, the $s_{\overline{x}}$. Going back to the previous example, it would be possible to repeat the measurements for $m = 31\ days$ on the sample of $n = 21\ mice$ [40].

The test does not give, as is often wrongly written, an evaluation of the means. It allows comparison of how much the measurements on single samples of mice deviate from the mean $\overline{x}_j$ and to what extent these means deviate from the mean of the means. It is the experimentalist who must decide what to ascribe these deviations to. The reason why a similar test is used, is precisely to see if there are effects that influence the measurement that can be identified as systematic errors. We must not believe, however, that the test can evaluate these effects. It can only give indications that there are such effects. Their numerical evaluation must be performed differently by studying the correlation between errors, repeating the measurements with variations of controlled parameters, etc.

Therefore the test gives only an indication if the standard deviations are caused by casual uncertainty. On the other hand, we already know that this uncertainty tends to zero as the number of measurements approaches infinity; the test also indirectly evaluates how far we are from

---

[40]In common usage it goes under the name of ANOVA test. It is a special case of the $\mathcal{F}$ test. In some texts this order is found inverted, and the $\mathcal{F}$ test is included as one of the ANOVA tests. It is important to have a clear idea of the meaning.

a sufficient number of measurements or if they are sufficient to draw conclusions from the experiment performed. This can be decided by fixing the value of the probability interval desired (see paragraph 4.9.1) or simply of the value of the probability from the area below the curve.

Coming back to the case presented, with the data in table 4.5, we obtain a value of $\mathcal{F} = 5.1$. For values of probability of 5% we have the value $\mathcal{F} = 1.93$, and for 1% we have $\mathcal{F} = 2.55$. The value we found is in any case higher. The probability of this occurring is $<< 1\%$; therefore, there is a significant difference between the two standard deviations. On the basis of the above-mentioned rules even the single value at 5% would have been sufficient to judge them significantly different.

Let us now see what this implies. This conclusion means that the standard deviation s_{xj} is significantly different from the one expected to make up $s_{\overline{x}}$. This may mean, for example, that something took place on a certain chosen day $j - th$ among the 31 of the observations (variations in the ensemble of mice, an ill mouse, or temperature variations that were not detected, etc.) such as to modify the behaviour of the distribution of the measurements on that day.

The student can compare these conclusions with those concerning the systematic error in the previous paragraph. In that case, the identification of the systematic error was relatively easier. In the latter, the experimental situation must be studied more carefully.

Let us now return to the need, or lack of need, for the Gauss distribution hypothesis. The conclusions we have reached are common in the practice known as the ANOVA test. They are based on the assumption that the standard deviations of a series of measurements tend to the same standard deviation for an infinite number of trials and this is valid also for the standard deviations of the means. This is tantamount to advancing the hypothesis that the distribution is of Gauss type or, from another viewpoint, that the theorem of the central limit holds. Therefore, on using the test, defining the parameter $\mathcal{F}$ with two standard deviations of the same data sample, we have to advance the hypothesis that the distribution is of the Gauss type; the test proves exactly this statement. In case of negative or doubtful answer, we cannot draw any other conclusion.

If the standard deviations are instead of two different and independent experimental distributions, then it is possible to compare them in a way similar to what we did in the example of the polarimeter. If, on the other hand, the process in discussion gives a distribution of experimental data with, for example, a Poisson parent distribution, then the test is still applicable, but in this case it would be meaningless to compare the standard deviations of the mean with the ones of a series of measurements.

We have discussed in detail the example of the ANOVA test to underline different aspects of the *applications of statistics in the biomedical field* in which it is widely used.

We wished especially to demonstrate that this is nothing more than an application of one of the different accuracy tests.

Furthermore, it was necessary to distinguish the fields of applicability, underlining the importance of keeping clearly in mind the hypothesis on the distribution of data.

It is also important to give students a general vision of the subject of the test of hypotheses, so that they are able to decide which to apply and how, instead of limiting themselves to rules learned by rote on examples that are not reproducible in their laboratory.

Coming back to the subject of the applicability of the test, the conditions of repeatability of the measurement according to the standard rules must in all cases be borne in mind . It is meaningless, in fact, as pointed out several times, to look for systematic effects or draw other conclusions if the standard deviations do not belong to data distributions obtained in the conditions of *repeatability*, which we summarise here once again [41]:

identical measurement procedure;

----

[41] According to *I.S.O.* rules [1]

  same laboratory;

  same equipment;

  same experimentalist;

or otherwise of *reproducibility* [3], in which these are checked and
known with an uncertainty lower than the one evaluated in the final
result.

4.8   The generalised χ^2 accuracy test

  The parameters of the accuracy tests introduced so far are all based
on the assumption of the verification of maximum likelihood of distributions. Whether they are an experimental and parent distribution or two
series of similar measurements, the three methods presented depend on
the assumption that maximum likelihood exists, since the parameter t
and the parameter $\mathcal{F}$ depend on χ^2 defined as a parameter to minimise
in order to have maximum likelihood. We have seen that this is true for
Gauss distribution functions.

  It is possible to perform other accuracy tests for data distributions
whose parent distributions are not Gauss distribution functions.

  A frequently used case in the field of the biomedical sciences is the
parameter χ^2 defined in a similar way, but completely different in values
and use, though it has the same name as the one introduced for Gauss
distribution functions [42].

---

[42]Students are often confused by these different definitions, and rightly so, since
the definition of the parameter should be unique; it would be sufficient to call the
one discussed in this section, for example, b^2! The justification is plausible, since the
Binomial distribution approximates a Gauss one for $\zeta \ll 1, n \to \infty, \mu \to \infty$; under
these conditions, the χ^2 that is introduced now tends towards the one introduced for
the least squares method.

For any distribution function the parameter

$$\chi^2 = \frac{(x - \mu)^2}{\sigma^2} \tag{4.148}$$

is defined. For a distribution parent to a Binomial distribution [43] we have $\mu = n\zeta, \sigma^2 = n\zeta(1 - \zeta) \equiv n\zeta\iota, x_\zeta = successes, x_\iota = fails.$ Therefore

$$\chi_b^2 = \frac{(x_\zeta - n\zeta)^2}{n\zeta\iota} \tag{4.149}$$

This is a general expression. We shall now see how to obtain a frequently used one and how to apply it.

---

[43] As will be remembered, the Binomial distribution function was introduced as the first and most general one; from this we obtained the other noteworthy distributions.

We rewrite equation 4.149:

$$\chi_b^2 = \frac{(x_\zeta - n\zeta)^2}{n\zeta\iota} = \frac{(x_\zeta - n\zeta)^2}{n\zeta(1-\zeta)} =$$

$$= \frac{(x_\zeta - n\zeta)^2}{n\zeta(1-\zeta)}(\zeta - \zeta + 1) =$$

$$= \frac{(x_\zeta - n\zeta)^2(1-\zeta)}{n\zeta(1-\zeta)} + \frac{(x_\zeta - n\zeta)^2\zeta}{n\zeta(1-\zeta)} =$$

$$= \frac{(x_\zeta - n\zeta)^2}{n\zeta} + \frac{(x_\zeta - n\zeta)^2}{n(1-\zeta)} =$$

$$= \frac{(x_\zeta - n\zeta)^2}{n\zeta} + \frac{(n\zeta - x_\zeta)^2}{n(1-\zeta)} =$$

$$= \frac{(x_\zeta - n\zeta)^2}{n\zeta} + \frac{(n\zeta - x_\zeta - n + n)^2}{n(1-\zeta)} =$$

$$= \frac{(x_\zeta - n\zeta)^2}{n\zeta} + \frac{(n - x_\zeta - n(1-\zeta))^2}{n(1-\zeta)} =$$

$$= \frac{(x_\zeta - n\zeta)^2}{n\zeta} + \frac{(x_\iota - n\iota)^2}{n(1-\zeta)} \qquad (4.150)$$

In this expression the dependence of parameter χ^2 on the number of successes and the number of failures is explicit. These are two different classes of data, thus they can be deduced from the general expression, simply by remembering that the χ^2 for ensembles of independent data can be added [44]. We can therefore see the number of successes as a sample of data, that of failures as another sample of independent data. Therefore expression 4.150, can be referred back to

$$\chi^2 = \frac{(x_1 - \mu_1)^2}{\sigma_1^2} + \frac{(x_2 - \mu_2)^2}{\sigma_2^2} \qquad (4.151)$$

---

[44]This property has not been explicitly deduced, but it is sufficient to think of the probability functions $\wp(\chi^2)$, which can be added for samples of independent measurements. The same also applies to the variable χ^2.

This is to say to the general expression of equation 4.148, from which, in fact, it was deduced. It is important to note that on *generalising the definition* 4.148, this can also be applied to a single value to evaluate how this is far from the mean value μ of the parent distribution, with standard deviation σ. It is necessary, however, to verify the existence of a parent distribution with a certain μ, and, in this specific case, if the parent distribution is Binomial. These verifications are often omitted when the χ^2 test is applied to a specific case. This vanifies the test of the hypothesis, which becomes meaningless. Let us examine a case in which the above-mentioned conditions are verified.

Exercise: *probability of being born black and born white*

Let us consider the population of a town in which there are exactly $\frac{3}{4}$ of black men or women and $\frac{1}{4}$ of white men or women [45].

In one day only 59 blacks and 14 whites are born regardless of sex. Is this event plausible?

Let us calculate parameter

$$\chi^2_{or} = \frac{(x_\zeta - n\zeta)^2}{n\zeta} + \frac{(x_\iota - n\iota)^2}{n\iota} = \tag{4.152}$$

$$= \frac{(59 - 73\frac{3}{4})^2}{73\frac{3}{4}} + \frac{(14 - 73\frac{1}{4})^2}{73\frac{1}{4}} = 1.3 \tag{4.153}$$

Determine $P(\chi^2 \geq 1.3)$, bearing in mind that the number of trials is not 73 but 1, because there is one measurement in one day. Therefore the number of degrees of freedom of the sample of data (which is 1, being a detection) is $\nu = 1$.

---

[45] It is interesting to note, in this example, that the number of existing individuals, namely the sample of experimental data, is taken as a reference to determine exactly the probability of successes and failures. This determines the parent distribution itself (Binomial).

Exercise: *probability of being born black or white, male or female*

We often find a similar exercise but with the reverse procedure. Here we can use the test to evaluate x_ζ and x_ι if we rely on the χ^2 obtained.

Repeat the calculation similar to the one in the previous exercise, to evaluate how many females were born that day, knowing that in the town there are $\frac{3}{4}$ females, $\frac{1}{4}$ males.

4.9 *Confidence, trust* and expanded uncertainty

To have a better idea of the quality of conclusions in the evaluation of hypotheses, here we shall examine some applications of accuracy tests. At the end of the paragraph we wish to give a rigorous definition of the uncertainty to associate with the value of the measurement in the notation. This is a question we raised in chapter 1.2.1, to which we can give an answer only now.

To do this in a complete manner, it would be necessary to discuss in greater depth the concepts and definitions of probability theory. Though fascinating, it is not possible to go into detail in this field here. We give only some hints in order to be able to use the concept of *confidence levels*, at least in certain cases.

In the most common definition used in texts on applied statistics, only the *frequentist* definition of probability is mentioned. By this is meant a limit on the number of trials repeated to ∞ of the *frequency* of an event (intended as in the example of the radioactive decay experiment).

The definition given above is not the only one. There are other approaches that introduce other evaluations in the definition of probability. For example, the *degree or level of trust* in a certain occurrence rather than in another, which can in any case be found in a certain population and which can be rigorously defined, although it is subjective. This is the field of Bayes statistics, of which there are several good introductions [46], which we recommend.

There are also other definitions and more general theories, but the two approaches mentioned here are the most common and they are connected with each other.

In what follows we shall attempt to give a general definition of the intervals of probability without an a priori definition of the probability concept. It will be enough for the student to retain the idea introduced in chapter 2.2 as intuitive of the probability concept. We will thus be able to arrive at the concept of *level* of *confidence* or *trust*.

4.9.1 Probability intervals

Let us consider a Gauss-like experimental distribution of measurements. We shall see in the following how to generalise the considerations for application in other distributions.

We know that in the interval between $\mu - \sigma$ and $\mu + \sigma$, a measurement has about 68% chance of being obtained. We call this the *interval of probability*. We have also seen other intervals of probability, for 2σ, 3σ etc. For coherence, we should say that the intervals are for $1 \cdot s$, $2 \cdot s$, $3 \cdot s$, etc., assuming that the approximation of the experimental distribution to a Gauss parent distribution holds.

We can write that the measurement has a value estimated between the intervals

$$\overline{x} - k \cdot s < x < \overline{x} + k \cdot s \tag{4.154}$$

that is

$$\overline{x} \pm k \cdot s \tag{4.155}$$

---

[46]See for example the complete introduction in the references [4] and the cited works.

with k called *coverage factor*.

Therefore k is a factor which, on varying, can *extend* the probability interval in which the measurement has to be considered.

We use k to characterise the measurement itself. If it is possible to associate a wide interval with a wide coverage factor with the measurement, the probability that the measurement falls in this interval increases with the reliability of the measurement. In fact, equation 4.154, can be rewritten as:

$$x - k \cdot s < \overline{x} < x + k \cdot s \tag{4.156}$$

where we indicated the estimate of the true value as included in the interval of probability with $\overline{x}$.

We shall now see how to proceed in practice. We usually have the measured quantity x and its estimate from repeated measurements, $\overline{x}$. We verify the type of distribution, for example Gauss-like (using, for example, the χ^2 test). We fix the coverage factor (which can be done by general rules) and we determine the width of the probability interval. The corresponding value gives the probability that a measured value includes the best estimate, $\overline{x}$, which is the mean value itself μ.

For example, let us take a Gauss distribution of 20 measurements. Bearing in mind that we can rely on a coverage factor $k = 2$, the interval of probability is extended between $\overline{x} - 2s$ and $\overline{x} + 2s$. That is, we have a 95.45% probability that the measurements falling within the interval $x - 2s, x + 2s$ include the estimated value $\overline{x}$. If there are 20 values, 1 of them will certainly not correspond to the true value, whilst the other 19 will all have the same probability of coinciding with the true value.

4.9.2 The *confidence* or *trust* level

The above explanation may appear laborious. It shows only in part the difficulty of the issue, which here is only touched upon. We must remember that here we need only to determine rigorously the expanded uncertainty to associate with the measurement. Let us then go back to probability values and the coverage factor.

The value of the integral in the interval of probability, is called the *confidence level*. The similarity with the common definition is obvious, since the confidence level that we assign to a measurement increases with its probability, which is exactly what we wish to represent [47]. We also speak about the level of *trust*, when we wish to distinguish a different concept of probability, such as, for example, in the case of subjective probability. However, as we have stated previously , this changes the meaning of *level* just in the content, but not in the way to apply it and to write it, to which we are now limiting our consideration.

If the distribution is a Gauss one, then we know that there is a correspondence between coverage factor and probability, according to the explanation given in chapter 2.2.3. For example, for $k = 1$, the level of confidence is 68.27%, for $k = 2.576$ it is 99.00% and so on.

For the presentation of the experimental result with extended associated uncertainty, in practice, it is of common use to combine the information on the combined uncertainty with that of the confidence level.

Let us consider the combined uncertainty obtained as described in chapter 3.1 from the combination of the standard deviations of different distributions. If we had not introduced the concept of confidence level, we could have written, for a quantity x (assuming for simplicity's sake that this is also the best estimate), whose combined uncertainty, as the experimental standard deviation from other distributions, is s_c, the

---

[47] In textbooks we often find confusion between probability intervals and *confidence intervals* which here are not defined, but they do correspond with the latter in numerical value but not in the extremes. This is an ambiguous definition and in fact it has been abolished internationally, according to [1] *I.S.O.* rules, but it is often found when explaining the principle of application of the confidence level. It is also good not to confuse it with another one, the *confidence interval*: we often speak about the *level of confidence* instead of confidence level or confidence interval.

expression with associated uncertainty:

$$x \pm s_c \tag{4.157}$$

The expression is similar to the one in chapter 1.2.1. We shall see in the following paragraph how to include in this expression also that of expanded uncertainty thanks to the confidence level.

4.9.3   The t parameter and the confidence level

Let us return to equation 4.156 for the values of x in the confidence interval, with best estimate $\overline{x}$:

$$x - ks < \overline{x} < x + ks \tag{4.158}$$

We can write for the parent distribution:

$$x - k\sigma < \mu < x + k\sigma \tag{4.159}$$

Generalising to the expression for the mean and the uncertainty of the mean, in the simple case of n distributions with equal experimental standard deviation s, parent μ:

$$\overline{x} - k\frac{\sigma}{\sqrt{n}} < \mu < \overline{x} + k\frac{\sigma}{\sqrt{n}} \tag{4.160}$$

which is similar to equation 4.159 for only one distribution. Rewriting equation 4.160 for the experimental distributions with equal standard deviation s:

$$\overline{x} - k\frac{s}{\sqrt{n}} < \mu < \overline{x} + k\frac{s}{\sqrt{n}} \tag{4.161}$$

This can also be written as:

$$\frac{ks}{\sqrt{n}} > |\overline{x} - \mu| \tag{4.162}$$

namely

$$k > \left| \frac{\overline{x} - \mu}{\frac{s}{\sqrt{n}}} \right| = \left| \frac{\overline{x} - \mu}{s_{\overline{x}}} \right| \tag{4.163}$$

Coming back now to the definition of parameter t:

$$t = \frac{\overline{x}}{s_{\overline{x}}} \tag{4.164}$$

which, by adding a constant equal to the mean value of the parent distribution, becomes:

$$t_\mu = \frac{\overline{x} - \mu}{s_{\overline{x}}} = \frac{\overline{x} - \mu}{\frac{s}{\sqrt{n}}} \tag{4.165}$$

which is precisely the second part of equation 4.163, which can also be written:

$$k > |t_\mu| \tag{4.166}$$

But it also holds that

$$t > t_\mu \tag{4.167}$$

Therefore we can rewrite equation 4.163 as

$$t > \left| \frac{\overline{x} - \mu}{\frac{s}{\sqrt{n}}} \right| \tag{4.168}$$

or in the extended form:

$$\overline{x} - t s_{\overline{x}} < \mu < \overline{x} + t s_{\overline{x}} \tag{4.169}$$

Therefore, parameter t replaces coverage factor k, or better still, we can identify the latter with t, of which we know the probability distribution function. We furthermore know that the distribution closely depends on the number of degrees of freedom.   Therefore the confidence level, determined with the interval of probability in equation 4.169, depends on the number of degrees of freedom. This makes the definition of level more direct and univocal.

Let us summarise what we have learned so far. The confidence interval depends on parameter t, whose distribution, similar to the Gauss one, gives the values of probability in correspondence to the degrees of freedom. We can determine k, which is to say, t of the probability interval with greater accuracy. We only have to define which probability value we think is acceptable, as we did previously in the case of the use of the parameter for a series of measurements.

Let us suppose, for example, that we have a series of measurements with $\nu = 13$ degrees of freedom and $\bar{x} = 50.00$, $s \cong 2.18$, $n = 16$, having fixed a value of confidence level 95.0%, then $t = 1.771$ (from tables). Therefore we can write equation 4.169

$$50.00 - 1.77 \cdot \frac{2.18}{\sqrt{16}} < \mu < 50.00 + 1.77 \cdot \frac{2.18}{\sqrt{16}} \qquad (4.170)$$

Namely

$$49.03 < \mu < 50.97 \qquad (4.171)$$

or

$$\mu - 0.97 < \bar{x} < \mu + 0.97 \qquad (4.172)$$

that is

$$\mu - 0.97 < 50.00 < \mu + 0.97 \qquad (4.173)$$

This expression leads us exactly to where we wished to arrive: we can define the expanded uncertainty $S = 0.95$, which can also be written as

$$\mu \cong \bar{x} \pm S = (50.00 \pm 0.97) \qquad (4.174)$$

It is to be noted that this is the best estimate of the true value; it contains expanded uncertainty S.

Students should not confuse the expression of expanded uncertainty $ks_{\bar{x}}$ with $s_{\bar{x}}$ itself. It is therefore best to avoid the notation $\Delta\bar{x}$ and not to identify it with expanded uncertainty, keeping the definition $\Delta\bar{x}$ as $s_{\bar{x}}$ given in chapter 3.6, weighted and not. The same applies for $\Delta\mu$, meant as $\Delta\bar{x} \cong \Delta\mu$.

Equation 4.174, is similar to 1.14 which we wrote in chapter 1.2.1.

In general we can state that t is the coefficient that multiplies not only the experimental standard deviation of the mean, but also the combined uncertainty, however obtained. This statement holds whether it is the combination of several experimental uncertainties, such as the reading resolution combined with the experimental deviation of a single distribution of data, and so on, or the standard deviation of the mean. Therefore, expanded uncertainty (which for simplicity's sake we can indicate with capital S) will contain information on instrumental errors, information on the single distributions which contribute to the final result, the procedure with which to combine them and the confidence level. For this reason, indicating the generic units of measurement with um, , the expression

$$(x \pm S)um \tag{4.175}$$

is the one widely used, since it is highly rigorous and complete.

This expression is more general than the one we have used so far, with which is often indicated the uncertainty due to the reading error, often identified directly with the resolution of the measurement instrument (for example Δl_i). To find in particular the correspondence between expanded uncertainty and reading uncertainty, we can consider the distribution function of constant probability seen in chapter 2.2.5. In this case of a single reading, we have $\nu = 1$ degrees of freedom, with a confidence level of 95.45%: $t = 13.97$. The interval is therefore

$$x - 13.97\frac{s/\sqrt{12}}{\sqrt{1}} < \mu < x + 13.97\frac{s/\sqrt{12}}{\sqrt{1}} \tag{4.176}$$

In the case of the example in chapter 2.2.5, with $\sigma \cong s = 0.1mA/\sqrt{12}$ namely $s = 0.03mA$ and value $x = 0.51$, we have:

$$0.51 - 13.97 \cdot 0.03 < \mu < 0.51 + 13.97 \cdot 0.03 \tag{4.177}$$

that is

$$(x \pm S)um = (0.51 \pm 0.09)mA \tag{4.178}$$

Therefore, for the reading of a single datum on a scale, since it is known only the reading uncertainty, estimated equal to the interval between two marks, the expanded uncertainty is approximately the interval between them. If a lower confidence level is required, for example 68.27%, then

$$(x \pm S)um = (0.51 \pm 0.06)mA \tag{4.179}$$

That is to say, the expanded uncertainty is lower but so is *the level of trust*.

In practice, this procedure is often omitted for a single reading. The procedure is justified if we have a number of measurements such that their distribution tends towards a Gauss one. It is therefore acceptable simply to indicate the value read with a certain reading uncertainty, in line with what is described in chapter 2.2.5. This example, however, is useful in showing that the estimate $s/\sqrt{12}$ represents a lower limit for the reading uncertainty. In fact, even for the 95.45% confidence level, the expanded uncertainty $(0.09mA)$ is larger than the estimate $s/\sqrt{12}$ $(0.03mA)$, which we would have for confidence level 50%, therefore unjustified. The caution in the use of relation $s/\sqrt{12}$ recommended previously in chapter 2.2.5, is now even more legitimate.

4.10 Exclusion of data

It is often the case that among the experimental measurements some are particularly doubtful. This may appear to be in contradiction with what has been stated thus far, in particular in the last paragraph, but in reality it is precisely an application of what we saw in 4.9.3.

Suppose that six students have each made a measurement of the length of a used pencil with the same ruler. Among the results, given in table 4.6, l_5 has a value clearly quite different from the others. We will

l_1	7.6cm	±0.1cm
l_2	7.0cm	±0.1cm
l_3	7.8cm	±0.1cm
l_4	6.8cm	±0.1cm
l_5	3.6cm	±0.1cm
l_6	7.8cm	±0.1cm

Table 4.6: Length of a used pencil measured by six different students.

keep it for different reasons. First of all because, with respect to the so called resolution or reading uncertainty of $0.1cm$, the datum $3.6cm$ is several times $0.1cm$ to the next values, and there are no other data with comparable difference.

Let us try to quantify these statements more precisely. The arithmetical mean is

$$\bar{l} = \frac{1}{6} \sum_{i=1}^{6} l_i = 6.8cm \qquad (4.180)$$

It is to be noted that this value contains the very measurement on which we entertain doubts. This means that the value is biased, i.e. it has some a priori condition of miss-trust.

The mean value found gives further reason to doubt the $3.6cm$ measurement since, even with respect to the arithmetical mean as can also be seen in figure 4.15.

To quantify better what we stated, we calculate the experimental standard deviation.

$$s = \sqrt{\frac{1}{5} \sum_{i=1}^{6} (l_i - \bar{l})^2} = 1.6cm \qquad (4.181)$$

The previous evaluation on the basis of reading resolution thus loses meaning. In fact, the difference of the value of $3.6cm$ turns out not to be as big in terms of standard deviation. On the other hand, the *doubtful* datum is $2 \cdot s$ times different from the arithmetical mean (see also figure 4.15):

$$l_5 - \bar{l} = 3.2cm \equiv 2 \cdot s \qquad (4.182)$$

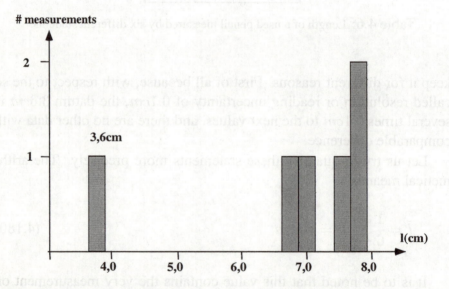

Figure 4.15: Data distribution for six measurements of length of a bar. The value to be evaluated for confidence is $3.6cm$.

We have not yet explicitly formulated the hypothesis that the distribution is Gauss-like. However, we have advanced some considerations that assumed that it was or should be such. In fact, given the phenomenon under examination, there is no reason why the distribution should not be Gauss one for a high number of measurements. But this is not the case now and, in fact, this is intrinsic to our difficulty in considering the data as valid or not. If we had a very large number of measurements we could state more confidently that the hypothesis of the datum is due to erroneous measurement or not. Therefore the previous evaluations on the difference between the datum and the arithmetical mean are ungrounded, even though these are the first ones that come to mind. This confirms what we stated on the difference of $2 \cdot s$.

To advance more quantitative and rigorous considerations, we must define the confidence interval within which to evaluate the probability of the datum in question.

The considerations in paragraph 4.9 were given for any distribution. Here it is necessary to hypothesize that the distribution function is Gauss-like, otherwise we could not determine any confidence level, since we could not define a probability interval. We will return later to the possibility of defining confidence levels for other distributions in a later paragraph about limits.

Let us suppose, only to make this exercise work, that there are no such systematic errors to deform or in any case modify the distribution function from a simple Gauss one. Let us also suppose that the procedure is simply the one in discussion and that there are no other phenomena that could lead to different distributions of other types (Poisson or others). Therefore the only limitation present in the evaluation of the confidence level is the fact that the number of events, or measurements, is low. In fact, we do not know if for $n \to \infty$ the distribution function that best approximates the data is a Gauss one. However, if we wish to draw conclusions in this experimental situation, it is necessary to hypothesize that the distribution is of Gauss type.

With this premise we can determine the value of the confidence level which, as we saw in the previous paragraph, depends on the parameter t. This parameter has a distribution, similar to the Gauss one, and its values of probability, put in tables in correspondence to the degrees of freedom, allow us to determine coverage factor k. We must simply define what we consider an acceptable value for probability, as we did in the case of the use of the parameter for a series of measurements. Let it be 92% with a coverage factor of $k = 2$; taking into account only the experimental standard deviation of uncertainty, the measurement $3.6cm$ has a probability of occurring (namely that it as correct as those of the other students) of about 8%.

To evaluate if it is acceptable or not, it is necessary to follow a conventional criterion, with the widest possible consensus. For this reason it is advisable to follow the indications given by *I.S.O.* in [1], which are the most common in practice. A confidence level of 95% is considered easily acceptable and in this case we shall consider the measurement in discussion acceptable as well.

Let us now go back to what we said above about the coverage factor. We can hypothesize that the experimental standard deviation is not the only uncertainty to associate with the measurements. It is therefore necessary to determine expanded uncertainty by determining coverage factor k, which will not be exactly equal to 2 as we assumed before. To do this we use the parameter t.

The number of measurements is $n = 6$, the number of degrees of freedom is $\nu = 6 - 1 = 5$, the chosen conventional value of the probability interval is 95% and from tables $t = 2.015$. We can now apply equation 4.169, which gives the probability interval for the true value μ:

$$\bar{x} - ts_{\bar{x}} < \mu < \bar{x} + ts_{\bar{x}};$$

$$6.8cm - 2.015\frac{1.6cm}{\sqrt{6}} < \mu < 6.8cm + 2.015\frac{1.6cm}{\sqrt{6}};$$

$$5.484cm < \mu < 8.116cm$$

which, considering reading uncertainty, approximates to

$$5.5cm < \mu < 8.1cm \tag{4.183}$$

This means that the true value of the measurement falls between $5.5cm$

and $8.1 cm$ *within the interval of probability* 95% (or better *with a confidence level* of 95%).

What can we thus conclude about the measurement of $3.6 cm$? Since it does not fall in this interval, it is not plausible as the true value, not even by repeating the measurements many more times, *therefore the measurement has to be rejected.*

It is to be noted that this is opposite to the conclusion we reached based only on the experimental standard deviation. Since the second approach is more rigorous, complete and reliable, we believe that the final conclusion, that *the measurement should be rejected*, is more reliable.

Let us now suppose that we repeat the same line of reasoning but with the conventional probability value in the interval equal to 99% instead of 95%. The value of the coverage factor is thus now equal to $k = t = 3.365$. That is, the true value is between

$$\overline{x} - t s_{\overline{x}} < \mu < \overline{x} + t s_{\overline{x}};$$
$$6.8 cm - 3.365 \frac{1.6 cm}{\sqrt{6}} < \mu < 6.8 cm + 3.365 \frac{1.6 cm}{\sqrt{6}};$$
$$4.602 cm < \mu < 8.998 cm$$

which, again taking into account reading uncertainty, approximates to

$$4.6 cm < \mu < 9.0 cm \tag{4.184}$$

The conclusion is the same for this value of the probability interval as well: since the interval of the true value does not cover the datum, *the measurement has to be rejected.*

In this paragraph we presented an application of probability intervals with an (*inverted*) logic with respect to the reasoning on the basis of which we introduced probability intervals. In fact, the wider the interval, the more restrictive, but more reliable, should our conclusions be. In the case of the rejection of data, instead, a wider interval, for example 99.9%, would lead to the conclusion that the measurement is plausible, since we obtain an interval of

$$2.3cm < \mu < 11.3cm \tag{4.185}$$

In reality, this is to be interpreted as a possible measurement, but a very rare one. For example, in 1000 measurements, the event that one is equal to $2.3cm$ is not plausible. However, for $3.6cm$ and 1000 measurements, we can calculate the value of k such that the extreme of the interval includes this measurement. It is $k \cong 6.4$ with an interval of probability equal to about 99.85%. This is to say that in 1000 measurements one or two might coincide with it and, in 10000, there would be 5 [48]. In this case the event is interpreted as rarer and rarer.

4.10.1   Limits and exclusion of data

As concerns the case of the exclusion of data, we have seen how to interpret the value of the probability interval. If we wish to obtain limits on the value of a not directly measurable quantity, the case is different. For example, to state that the concentration of a certain substance, no trace of which has been detected, is lower than a certain measured value, it is necessary to have a good knowledge of the distribution, especially in the case in which it is symmetric. The extremes of the probability interval, in this case the lower one, are *the limits* as a function of which the limit value below a certain measured value at which the substance has not been detected, is determined. If the distribution is Gauss-like there is a linear relation between this limit and the extreme of the probability interval, otherwise the functions are more complicated and do not give symmetric limit values. As can be seen, the issue is complex, not lastly because it involves a review of the concepts of probability and must be dealt with in more advanced textbooks, which can be found in the references.

---

[48] It is interesting to note that for the considerations on experimental standard deviation alone, we would expect 5 events out of 100 instead of 5 out of 10000.

Appendix

A   Experiments

A.1   The speed of light in coaxial cables

To determine the speed of light in coaxial cables, we can use a very simple apparatus made of a double-trace fast oscilloscope (at least 200MHz), several pieces of coaxial cables of the *BNC* type and an alternating current generator, which gives two signals visualised on the oscilloscope, passing through two cables, one longer than the other.

Apparatus

200MHz Oscilloscope

pulse generator

carpenter's rule

BNC coaxial cables and adaptors

Procedure

The measurement consists of adding pieces of cable (for the signal arriving later on the oscilloscope) each time measuring each single piece.

The speed of light is given by the ratio between the total lengths d_x of the added cables and the time differences d_t between the two signals.

Analysis of measurement

Discuss the combination of instrumental uncertainty, combined uncertainty and mean uncertainty (see text).

Perform the χ^2 test for the relation between the lengths and the time intervals.

A.2 The light rotation angle in a polarimeter

By means of a commercial polarimeter it is possible to evaluate the systematic error in the measurement of the rotation angle of the polarisation plane of light and compare it with the specs resolution, thus determining the zero error of the scale.

Apparatus

commercial polarimeter equipped with a graduated scale and a vernier

thermometer (recommended)

Procedure

Measurement consists of the determination of the offset angle value with the same sample. The offset of the scale is defined as the position of equal shadow (same illumination) between the two half moons on the eyepiece. By rotating randomly and coming back to the same equal shadow image, we can perform the readings of the *offset* value which will be close to the offset of zero on the scale but will not be the same. Repeat this procedure for several measurements. Two different operators take two different series of readings on the same positions.

Analysis of the measurement

Evaluate the systematic uncertainty as the experimental standard deviation with the hypothesis of a Gauss distribution.

Evaluate the two series of measurements with the t test.

Apply the test of the $\mathcal{F}$ variable to the two series.

A.3   Radioactive decay counts

With a particle detector, a counter, a watch and if possible a radioactive source, it is possible to obtain different distributions of counts for different time intervals in a ionizing radiation detector.

Apparatus

particle detector (for example a plastic scintillator and a photomultiplier)

chronometer with a resolution at most of one tenth of a second

electronic counter

radioactive source (example ^{90}Sr)

If, for example, we use a plastic scintillator, then it is necessary to have a photomultiplier, a power supply, a discriminator and/or an amplifier with a digital electronic counter as well.

Procedure

Record a large number (200 or more) of counts in time intervals of of the same duration, for example $1s$. For the same total number of counts, take another series of measurements for a longer time interval, for example $5s$.

Analysis of the measurement

Perform the χ^2 test for both experimental distributions of the counts obtained, comparing them in both cases with the hypothesis of the Gauss function, and the Poisson function.

A.4   The time constant in an RC circuit

To determine the time constant τ of an RC circuit, we can assemble an elementary electrical circuit with a capacitor and one or more resistors and a switch. The circuit can be powered by a battery connected to the capacitor. We can measure the voltage at the output of the capacitor while it is discharging.

Apparatus

electrolytic capacitor (thousands of Farad);

three equal resistors (thousands of Ohm)

a switch

a battery

an analogue tester

a chronometer

Procedure

Charge the capacitor and then discharge it by opening the switch to let the current flow through the resistor. Measure the voltage at the leads of the capacitors at regular time intervals measured with the chronometer. Take the measurements up to an interval of time amounting to at least 3 times τ initially calculated as the theoretical value from the values of the resistors and the capacitor. Repeat the same series of measurements for one, two and three resistors in series.

Analysis of the measurement

Record the voltage values as a function of time.

Apply the χ^2 test with at least τ as a free parameter.

Repeat the same procedure for the three series of measurements, comparing them to calculated τ and to each other. Determine the weighted mean of constant τ of each series of measurements and of the calculated value.

A.5 Photoresistance of a semiconductor

We can study the behaviour of a simple commercial silicon photoresistor by the effect of incident light. We measure the variation of the resistor at the outlet of the photoresistor, which has to be properly aligned in an optical system with two polarisers, with direct light from a Sodium lamp.

Apparatus

lamp (for example a Sodium one)

polarising filters rotating on a graduated scale

Silicon photoresistor

tester

Procedure

Measure the values of the resistor at the outlet of the photoresistor for different rotation angle values in one of the two polarisers for the interval between 0° and 180°. The photoresistor can also be powered and in this case it will be necessary to have a suitable power supply. Measurements are to be made either with natural light or with artificial light in the laboratory as well as in conditions of maximum darkness.

Analysis of the measurement

Obtain the two resistor curves as a function of the angle.

Perform the χ^2 test for the hypothesis that the two curves both represent a Lorentz and a Gauss distribution function respectively.

A.6 Magnetic induction between two solenoids

We can study the induction of alternating voltage by inserting, for example, one solenoid inside a larger one. We can study the phenomenon for different voltage values, different solenoids suitably built with different numbers of windings, and different values of current and frequency. For suitable dimensions and number of windings it is possible to observe the resonance between the inducing and the induced voltage. The readings can be taken with an oscilloscope and a tester.

Apparatus

oscilloscope (a simple 20MHz one)

alternate voltage generator with variable frequency and amplitude

tester for alternating current measurements

solenoids (at least one large and three smaller ones to insert in the larger one)

Procedure

By inserting a solenoid (secondary circuit) inside a larger one (primary circuit), we can measure the voltage induced on the secondary one as a function of the voltages of the primary circuit at fixed voltage and frequency of the primary circuit. We repeat the same for different frequencies and currents, with a fixed primary voltage and all the solenoids we have. For suitable values of frequency, voltage and number of windings we can measure the increase in voltage as a function of frequency, as a resonance between the two circuits.

Analysis of the measurement

Perform the χ^2 test for the voltage and current curves of the secondary circuit as a function of those of the primary in the hypothesis of a linear function. With the χ^2 test, verify the hypothesis of an analytical function similar to the Lorentz distribution function for the voltage curves of the secondary as a function of the frequency of the primary, which show a behaviour similar to that of resonance.

Using a solenoid with an unknown number of windings, by extrapolation find the number of windings from the values of the voltage of the other solenoids with the same voltage and frequency as the primary circuit.

On this value determine the combined uncertainty due to extrapolation.

A.7   The characteristic curve of a semiconductor diode

To build the characteristic curves of a semiconductor diode it is sufficient to have a simple circuit with a commercial integrated Silicon (or also Germanium, which allows easier measurements in reverse polarisation) diode. We can study the behaviour both in reverse or direct polarisation, with great accuracy even in the zone of *ohmic* behaviour and determine contact potential and breaking point of the crystal.

Apparatus

commercial diode semiconductor (Silicon or Germanium)

power supply (for example, a simple 9 volt battery)

one or more potentiometers

testers (to check the charge of the battery and measure the current)

Procedure

Measure the current of the circuit as a function of the current at the ends of the diode, which is varied using the potentiometer connected in series to the battery. Measurements at low direct current must be accurate and frequent to determine contact potential. Reverse the polarity of the applied voltage and measure the reverse current. Depending on the quality of the diode, and possibly having aa second power supply to replace the battery, measure the breaking point of the crystal.

Analysis of the measurement

Perform the χ^2 test on the functions which approximate the experimental curves, for direct and reverse current, the former with a function of the exponential type and the latter with one of the linear type.

We can study the combination of the experimental uncertainties in the two cases: one in which we use the same tester to measure voltage and current and the other in which we use two different instruments. In the first case the combination of uncertainties will take into account the *correlation* between them, especially the systematic ones.

B Symbols used in the text

We list here the symbols most frequently used in the text, together with their definitions. Only some of them have two definitions: in most of these cases only the least obvious and most characteristic one is given.

α number of known parameters

β HWHM half width half maximum

γ ratio of differences in the extrapolation

δ interdistance between two marks on a graduated scale

$\delta...\mathbf{x}$ consecutive differences in interpolation

$\Delta\mathbf{x}$ interval of values equal to uncertainty

$\mathbf{d}$ deviation

$\mathbf{k}$ coverage factor

ε efficiency

η limit of the sum of the modulus of deviations

ξ variance

$\mathbf{q}$ degree of the polynomial

ϑ angle of rotation

μ mean value

$\overline{\mathbf{x}}$ arithmetic mean

$\mu_{\mathbf{w}}$ weighted mean value

$\overline{x}_{\mathbf{w}}$ weighted mean

ν degrees of freedom

χ^2 accuracy parameter

$\chi_{\mathbf{r}}^2$ reduced accuracy parameter

$\chi_{\mathbf{o}}^2$ observed accuracy parameter

$\mathcal{F}$ accuracy parameter

t accuracy parameter

σ parent standard deviation

σ_μ mean value standard deviation

$\sigma_{\mu_{\mathbf{w}}}$ weighted mean standard deviation

s experimental standard deviation

$s_{\overline{x}}$ mean standard deviation

$s_{\overline{x}_{\mathbf{w}}}$ weighted mean standard deviation

τ time constant of the RC circuit

List of Figures

List of Tables

Bibliography

In this first part are indicated books recommended to the students both for more advanced notions and for completeness of mathematical theorems.

[1] BIPM, IEC, IFCC, ISO, IUPAC, IUPAP, OIML, *Guide to the Expression of Uncertainty in Measurement*, International Organisation for Standardisation, Geneva (1995).

[2] R. Parncutt, *Harmony: A Psychoacoustical Approach*, Springer-Verlag, Heidelberg (1988).

[3] BIPM, IEC, IFCC, ISO, IUPAC, IUPAP, OIML, *International Vocabulary of basic and general terms in metrology*, International Organisation for Standardisation, Geneva (1993).

[4] G. D'Agostini, *Probability and Measurement Uncertainty in Physics - a Bayesian Primer*, Internal Note $n°1070$ 23 November 1995 INFN and Physics Department, University "La Sapienza", Rome Italy.

[5] R. J. Barlow, *Statistics*, John Wiley & Sons, Chichester (1989).

[6] P. R. Bevington, *Data Reduction and Error Analysis for the Physical Sciences*, McGraw-Hill Book, New York (1969).

[7] R. A. Fisher, *Methodi statistici ad uso dei ricercatori*, UTET, Torino (1948).

[8] A. G. Frodesen, O. Skjeggestad, H. Tfte, *Probability and Statistics in Particle Physics*, Universitetsforlaget, Bergen-Oslo-Troms (1987).

[9] T. H. Wonnacott, R. J. Wonnacott, *Introduzione alla statistica*, Angeli, Milano (1992).

> *In this second part are indicated the books that present a completely different analysis from the one reported in this textbook. Several parts are clearly in contrast with the definitions and discussions presented here.*

[10] R. L. Anderson, T. A. Bancroft, *Statistical theory in research*, Mac Graw Hill, New York (1952).

[11] M. C. Barford, *Experimental measurements: precision, error and truth*, John Wiley & Sons, Chichester (1985).

[12] D. C. Baird, *Experimentation*, Prentice Hall, Englewood Cliffs (3rd edition 1995).

[13] L. Cavalli Sforza, *Analisi statistica per medici e biologi*, Boringhieri, Torino (1961).

[14] M. A. Glantz, *Primer for biostatistics*, Mac Graw Hill, New York (1981).

[15] S. Brandt, *Statistical and Computational Methods in Data Analysis*, North-Holland, Amsterdam (1976).

[16] A. G. Kendall, *The Advanced Theory of Statistics, vol. I, vol. II*, Charles Griffin & CO. London (1952).

[17] W. R. Leo, *Experimental Techniques in Nuclear and Particle Physics*, Springer-Verlag, Berlin (1987).

[18] L. Lyons, *Statistics for nuclear and particle physicists*, Cambridge University Press, Cambridge (1986).

[19] J. Mandel, *The statistical analysis of experimental data*, Dover Publications, New York (1964).

[20] B. P. Roe, *Probability and Statistics in experimental Physics*, Springer-Verlag, New York (1992).

[21] J. R. Taylor, *An Introduction to Error Analysis*, University Science Books (1982).

[22] A. G. Worthing J. Geffner, *Treatment of experimental data*, John Wiley Sons, Inc. New York (1965).

[23] K. Wilson J. Walker, *Principles and Techniques of Practical Biochemistry*, Cambridge University Press, Cambridge (1994).

Index